A Quantum Theory of Color Strings: A Palette of Gluons

First Edition

Dr. Robert Nieves

Library of Congress Control Number: 2022900093

ISBN: 9798795291154

© 2022 by Dr. Robert Nieves
All rights reserved

A Quantum Theory of Color Strings: A Palette of Gluons

This book is ideal for students, researchers, and readers in all areas of quantum physics, color string theory and color supersymmetry. A Quantum Theory of Color Strings probes the very limits of quantum physics and scientific possibility.

There are three main parts of the book focusing on the colors of the force, color supersymmetry, and the frontier of quantum physics, to discuss the latest theoretical and empirical findings in the quest for the veracity of physical reality.

This book explains in detail the fundamental color charge aspects that serve as a foundation to the Quantum Theory of Color Strings and Color Supersymmetry.

There are three principal ideas in this book. The first is that the colors of the force are the elementary color charges for the gluon standard model at the frontier of physics. The second idea is that gluons are the resulting combination of the fundamental color charges of the elementary particles of all there is. The color charge laws, or the color charge Maxwell equations, are the fundamental particle-wave equations of physical reality. The third idea is that six-dimensional space-time with a (3 + 3) formalism is emergent and fundamental to the color force and its charges.

Some of the questions that are addressed in this book, including, but not limited to: Are color strings fundamental perturbations with topological properties? What is the right number of gluons? What happens to an electron at parity? Could an electron represent a fundamental generation of gluon pairs with a negative charge due to parity? What is the wave function of a color string? Is there a color graviton? Why is the Higgs boson lighter than expected? Does the color graviton couple to the Higgs field? What is color loop quantum gravity? What is the significance of the Color Charge Maxwell Equations? Is a gluon a wave and a particle? What is the current status of Color Supersymmetry? Is it all a matter of waves and bosons? The author describes in detail these challenging subjects and how they impact physical reality.

Robert Nieves has a diversified professional experience in engineering, teaching, international business administration, and physics and cosmology research. Dr. Nieves holds a Bachelor of Science in Electrical Engineering from the Illinois Institute of Technology and an MBA and a DIBA from Nova Southeastern University in Florida.

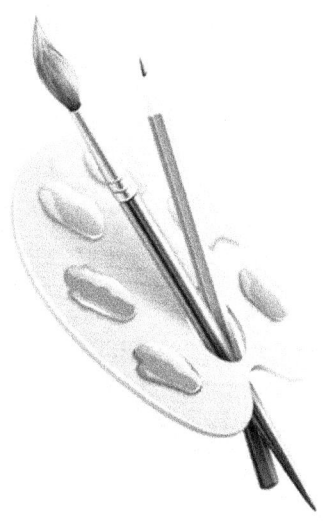

Dedicated to the creator of all there is

CONTENTS

PART I

THE COLORS OF THE FORCE

CHAPTER 1   1

The Quantum Theory of Color Strings

1. An introduction to color strings.
2. The four fundamental forces of nature.
3. The string colors of all things.
4. The strong nuclear force as the most potent force in nature.
5. Is a nucleon a Mealy deterministic finite-state machine?
6. So, what is the right number of gluons?
7. The color exchange mechanism.
8. The particles and anti-particles of the current standard model.
9. The Neutrino Problem of the Current Standard Model.

    9.1 The Peculiar Neutrino of the Gluon Standard Model.

    9.2 The Transmutation of Fermions by the Electroweak Force.

10. The gluon theory as the substructure of elementary particles.
11. The color string theory.
12. Are color strings fundamental perturbations with topological properties in the spatiotemporal fabric of physical reality?

CHAPTER 2   70

The Color Charge Maxwell Equations

1. The Laws of Color Charges.

2. What are the color charge maxwell equations in six-dimensional space-time?

3. What is the significance of the Color Charge Maxwell Equations?

4. What are the Color Charge Maxwell Equations?

PART II

COLOR SUPERSYMMETRY

CHAPTER 3	76

Color String Supersymmetry

1. The Theory of Color Supersymmetry (COSUSY).

2. The Potential of Color Symmetries.

3. The Theoretical Predictions of Color Supersymmetry.

4. A Concerto of Color Charges: Who is playing the color strings?

5. The Spinning Reflection of a Color Charge.

CHAPTER 4	95

The Spatiotemporal Unitary Color Group of Order "n" for Color Charges

1. Do Symmetries, Translations, Scaling, Reflections, and Rotations lead to the Discovery of the Laws of Nature?

2. What is a unitary group?

3. How does a finite unitary group appear in Quantum Mechanics prior to the Quantum Field Theory?

PART III

THE FRONTIER OF QUANTUM PHYSICS

CHAPTER 5   104

The Theory of Color Loop Quantum Gravity (CLQG)

1. Is the Color Spinor and its Color Spin Connection a case for quantum gravity with no color strings attached?

2. Could space-time be the source of both a dependent and an independent background of physical reality as well as the quintessential source of all there is?

3. Are the Higgs Scalar Boson and the Higgs Field part of the Color String Theory?

    3.1 The Relationship between Geometry and Energy Momentum for the Spatiotemporal Medium of Color Gravitons.

    3.2 The Hidden Thermodynamics of the Graviton and its Gravitational Wave.

    3.3 The Klein-Gordon Equation for an Effective Field Theory of the Color Graviton.

4. The Color Charge Field.

    4.1 The Unconfirmed Crucial Hurdle.

    4.2 The Quest for the Detection of the Color Graviton.

5. The Color Graviton Theory.

6. The Cosmological Graviton.

7. The Jiggle of a Quantum Gravitational Wave.

CHAPTER 6   150

The Lifetimes of Particles

1. Why does a muon have a slower decay than an electron?

2. Is it possible that the g–factor has been hiding in plain sight all along?

3. Could all the energies in our universe be added up into a single Lagrangian equation?

CHAPTER 7   179

The Accretion or Dissolution of Mass

1. What is the source and the sink of all color strings?

2. How does a color string sustain its energy density?

3. What is the quantum charge of a color string?

4. What makes the gluonic field stronger as color string charges separate? How do color string charges separate or spring back?

5. What is the angular momentum of a color string?

6. The Interference and the Geometry of a Cluster of Wavelets about a Point.

7. What is the color charge of a color string?

REFERENCES   199

PART I

THE COLORS OF THE FORCE

Chapter 1

The Quantum Theory of Color Strings

§ 1. An Introduction to Color Strings

It would probably be very surprising to masters like Velazquez, Goya, or Picasso, that science would try to imitate the visual beauty of the colors in their art with theories that deal so much with colors without having anything to do at all with any of the colors of paintings or the way that colors are perceived in our physical reality.

The combination of a red, a green and a blue quark, or the combination of a cyan, a magenta, and a yellow quark, is colorless. Quarks come in flavors. The different types of quarks, u for up, d for down, s for strange, c for charm, b for bottom, and t for top, is called the flavor of the quark, and color is a characteristic property that comes from the colorness of gluons. The colorness is defined as the color property that also has an inherent colorness potential.

The strongest force in our universe is the strong nuclear force. The strong nuclear force is present inside every nucleon of an atom. Each nucleon consist of quarks and each quark has its own color charge. If all three colors of the gluons inside a quark are combined the result combination is colorless by the laws of physics of our universe. The strong nuclear force has the most mystifying features of physical reality. Inside every nucleon, there are three color quarks, each quark has its distinctive color.

The colors inside all three quark combine into a colorless quantum state as directed by the laws of nature in our universe. There may be three quark combinations, three anti-quarks with their anti-colors, or quark-antiquark combinations with color-anticolor that offset each other. Recent research has also shown that there are tetraquarks with two quarks and two anti-quarks, and there also pentaquarks with four quarks and one anti-quark, to produce colorless quantum states.

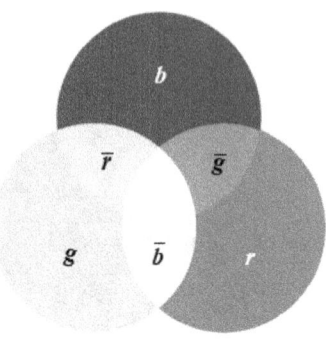

Figure 1. The Gluons of Color String Theory.

However, the gluon particles that mediate the strong nuclear force may combine into nine varieties of quarks if the photon is also included as a color singlet gluon that is colorless. The interactions of every fundamental calculation of the strong nuclear force in described by Feynman diagrams. The gluons are the quanta of the color force while the color charges of a gluon create a color gauge field. A color gauge is a coordinate system that varies with location with respect to some parameter or base space. A color gauge transform is a change of coordinates applied to a location, and a color gauge theory is the model of a mathematical or physical system to which a color gauge transform can be applied, and is typically color gauge invariant, in that all physically relevant color quantities are left unchanged or transform naturally under color gauge transformations.

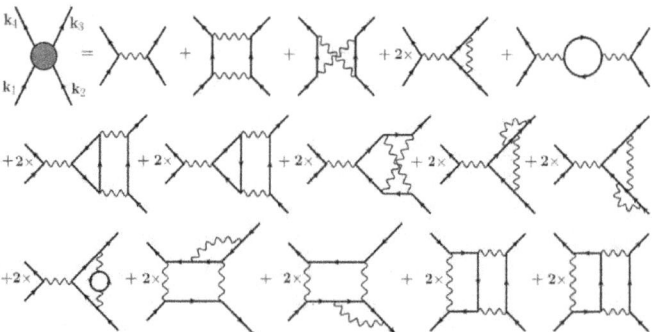

Figure 2. An Illustration of a variety of Feynman Diagrams used to describe the electromagnetic interactions mediated by a photon. (De Carvalho et Alia, 2013)

A Feynman diagram shows the trajectory of elementary particles when they collide. Feynman diagrams are very useful in quantum mechanics for the interactions of elementary particles. In Feynman diagrams, the particles may go forward or backward in time. An antiparticle is a particle going backward in time. The diagrams are applicable in Quantum Field Theory or in Quantum Gravity Theory like A Dynamic Theory of Space-Time.

The following are seven of the most common basic elements of Feynman diagrams:

Figure 3. Common Symbols for Feynman Diagram Elements.

Any Feynman diagram may be sectionalized to these basic elements of particle action. The diagrams may be classified as, but not limited to, lines of directional motion at an angle, particle-waves of emission, gluon curly waves, or loops of motion. The basic elements may be combined as into common geometrical shapes for the efficacy of the visualization of the interaction. Arrows for particles are left to right and for antiparticles are right to left. Space is up or down and time is left to right, unless otherwise noted. Gravity may be shown as a dashed line.

$\alpha \longrightarrow \beta \quad \rightarrow \quad \left(\dfrac{i}{\not{p} - m + i\varepsilon}\right)_{\beta\alpha}$

$\mu \sim\sim\sim\sim\sim \nu \quad \rightarrow \quad \dfrac{-i\eta_{\mu\nu}}{p^2 + i\varepsilon}$

$\begin{array}{c}\beta\\ \diagdown \\ \sim\sim\sim\mu\\ \diagup\\ \alpha\end{array} \quad \rightarrow \quad -ie\gamma^\mu_{\beta\alpha}(2\pi)^4\delta^{(4)}(p_1 + p_2 + p_3).$

Incoming fermion: $\alpha \longrightarrow\bullet \quad \rightarrow \quad u_\alpha(\vec{p}, s)$

Incoming antifermion: $\alpha \longleftarrow\bullet \quad \rightarrow \quad \bar{v}_\alpha(\vec{p}, s)$

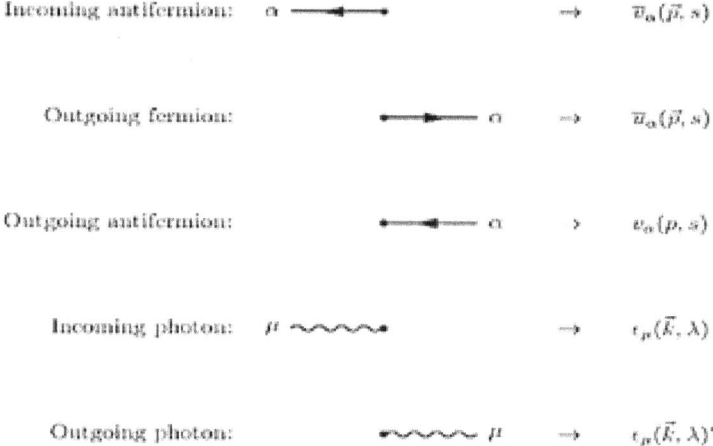

Figure 4. The rules for drawing the technique of Feynman Diagrams. (Peskin, 1995)

The frequency of the particle-waves may be constant during an interaction. Particle-waves are used for emission and interchange of exchange particles or gauge bosons. The circle may be some quantum interaction that may not be included yet for current calculations, and may be added later. A virtual electron-positron pair may be represented as a double circle. A continuous creation of pairs of virtual particles, as in vacuum polarization, may be represented as the circle with two arrows and particle-waves on each side. Handy tables have helped researchers and students to translate each Feynman diagram element to its corresponding mathematical expression.

For example, if two electrons approach each other, a virtual photon (a particle-wave) is exchanged between the two electrons causing them to repel. The closer the electrons approach each other, the shorter the virtual photon wavelength becomes.

Pair production may also be described as the result of parity on an arbitrary electron (outgoing arrow) by an incident photon (incident particle-waves) and a positron (incident arrow). The reverse of an arrow, the transposition of the other arrow, and the direction of the incident particle-waves with respect to the temporal axis, would describe an annihilation, an emission, or an absorption diagram. Each vertex would have an arrow in and an arrow out.

§ 2. The Four Fundamental Forces of Nature

It is theorized that there was a unified spatiotemporal force at very high energies that first divided into the present gravitational force, a force from the spatiotemporal geometry about mass or matter, and the grand unification force of color charges. After a period of time, the strong nuclear force emerged from the grand unification force of color charges, and much later the weak nuclear force and the electromagnetic force emerged from the electroweak unification force.

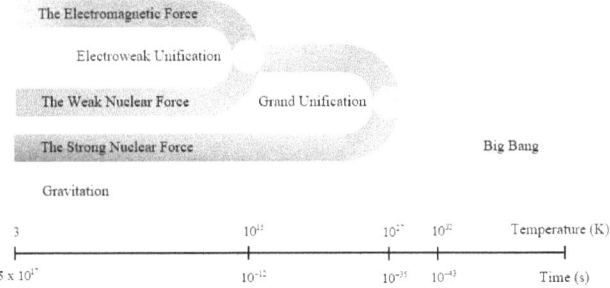

Figure 5. The Separation of the Four Fundamental Forces of Nature.

The Quantum Triadic Color Conjunction of color charge current "J^β", spatiotemporal curvature "Γ", and electrodynamics "E", constitutes the substructure of all four fundamental forces of nature. The gluon standard model emerges from the "$EJ\Gamma$": the strong nuclear force, the weak nuclear force, electromagnetism, charge, energy, mass, particles, force carriers, matter, and gravitational fields. (Nieves, 2020)

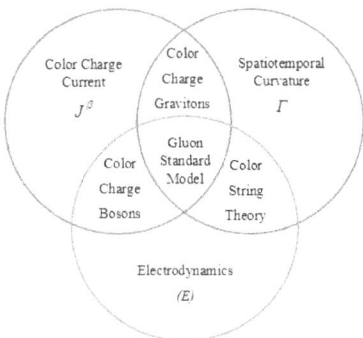

Figure 6. The Quantum Triadic Color Conjunction.

According to current theory, there are four fundamental forces of physical reality that are controlled by their own distinctive rules.

The fundamental force of electromagnetism provides a variety of charges, positive charges and negative charges, at a microscale of physical reality. Charges that are alike would repel and charges that are opposite would attract.

From previous research, electromagnetism and gravitation are both aspects of space-time.

The components of each fundamental force are free carriers of different magnitudes or strengths of each force without restrictions.

The photon for electromagnetism or the color graviton for gravitation mediate all possible interactions for each corresponding fundamental force field.

A force carrier, or gauge boson, is a mediatory particle that carries any of the fundamental interactions. Elementary particles interact with each other by the exchange of bosons.

Interaction	Boson	Relative Strength	~Mass (eV/c^2)
Strong	gluons	1	0
Electromagnetism	photon	10^{-2}	0
Weak	W^+, W^-, Z^0	10^{-13}	80 G, 80 G, 91 G
Quantum Gravitation	Color Graviton	10^{-39}	10^{-24}

Figure 7. Fundamental Interactions and Force Carriers

In a gravitational field, there is a quantum gravitational charge, the color graviton or the mass of a hypothetical singularity, that may be either attractive or repulsive depending on the angular frequency of the color graviton, the property of the mass; that is, the property of the colorness of the quantum constituents of that mass, and whether the corresponding space-time is expanding or contracting.

For example, the gravitational field of a black hole is theorized to be attractive and the gravitational field of its corresponding white hole is theorized to be repulsive.

As mass contracts, it is theorized that it may create a black hole-white hole pair that obeys a quantum theory of gravity. Gravitation is the result of the geometry of space-time around a color graviton, a quark, a hadron, an elementary particle, or a system of particles at any scale of physical reality.

Gravitation may occur at the quantum scale or at a macroscale of physical reality as long as the conditions are present for the manifestation of a gravitational field.

The graviton may carry gravitational charge because of its constituency and its geometry within the fabric of space-time. However, the color graviton is not theorized to be the only and single manifestation of gravity. Curved space-time by itself can be a gravitational field wave.

§ 3. The string colors of all things.

The success of a theory of color strings would be a convergence of the beautiful and truthful ideas in modern physics, each of which feels correct and verifiable in their own unique way.

It is also inevitable that there must be a more beautiful, truthful, and correct understanding of space-time. Space-Time is not a vacuum; Space-Time is a volume.

Space may be considered a relation, an entity, or a quintessence. The content of space is dependent on the entity of space. Space may be considered an entity if space would exist without physical objects, or a relation if space would consist entirely of quantum fields in the contiguous spatial medium. So, how is space or time a relation? The early debate through correspondence between the physicist Isaac Newton and the mathematician Gottfried Leibniz accentuates two views of time, Leibniz had the view that there is only relative time and that space or motion exists only as a relation between objects, while Newton proposed that the motion of objects occurred in relation to an absolute and independent frame of reference from the objects therein contained. These two ways of thinking about space and time offer a functional description of space-time as the quintessence of physical reality. From previous research, it was theorized that time may have both an absolute aspect and a relative aspect in its wave property. (Nieves, 2021)

Space-Time is a complex field. Space-Time consists of points which are discrete by definition, where a point may or may not have a local event. The collection of the quanta of space points and their events can represent a quantum gravitational space-time. Any spatiotemporal point source may spin as space endows more time, and time endows more space, since the expansion or contraction of space-time has a wave property. As long as the spatiotemporal field expands or contracts, time will pass in the quantum gravitational space.

Space, time, particles and their fields (quantum fields) constitute covariant fields. Our physical reality may be described by relations, dependent entities with seemingly independent behavior, the quintessence of space, and wave property. The constituents of physical reality are quantum fields. Any entity that relates to, influences or it is influenced by another, exists in the history of our physical reality, even if the entity originated outside of our universe. Our universal physical reality is the network of all interacting quantum fields and their events within our universal and contiguous space-time.

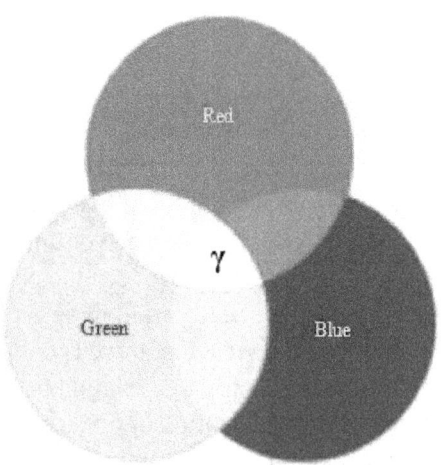

Figure 8a. An Illustration of the additive color combination of a colorless Photon.

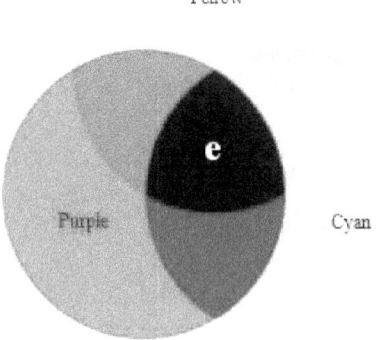

Figure 8b. An Illustration of the additive color combination of an electron due to parity.

The rules of the strong nuclear force are based on three fundamental color string charges and obey different rules than other fundamental forces in nature. These rules are given as follows:

- There are no net charges of any color after color string charges combine, only colorless states are allowed.

- A color string pair, a color and its anti-color is colorless, or the combination of three color and three anti-color pairs is colorless.

- A gluon has a color string charge of one color. String color charges may be red, anti-red, blue, anti-blue, green, and anti-green.

- Each quark or each anti-quark has a net color string charge of one color or one anti-color. Quarks and anti-quarks may exchange color gluons to form their bound states.

These complex rules of the strong nuclear force explain how nucleons are held together. Nucleons like protons and neutrons are made of quarks and quarks are made of gluons, gluons consist of color string charges. Protons, neutrons and baryons are all made of three quarks, each quark has a distinct color string charge. Every particle, including each proton or neutron has an anti-particle that is made of three anti-quarks, each having a distinct color string charge.

Nonetheless, every combination of color string charges that make up a particle or an anti-particle at any time has to be colorless. Thus, a color must bind with an anti-color in the allowable combinations that exist in nature.

§ 4. The Strong Nuclear Force as the Most Potent Force in Nature

It is theorized that the strong nuclear force of color string theory is manifested through the exchange of color charges or gluons. As the strong nuclear force emerges from the gluonic field, the color charges or anticolor charges of the quarks or antiquarks change. The exchange of color charges between quarks or antiquarks is unique to the string force.

Gluon Pair		Quark		
		Present State	Next State	Action
Color	Anticolor	Q_n	Q_{n+1}	
Red	Yellow	Red	Blue	Emission
Red	Yellow	Blue	Red	Absorption
Red	Purple	Red	Green	Emission
Red	Purple	Green	Red	Absorption
Blue	Cyan	Blue	Red	Emission
Blue	Cyan	Red	Blue	Absorption
Blue	Purple	Blue	Green	Emission
Blue	Purple	Green	Blue	Absorption
Green	Cyan	Green	Red	Emission
Green	Cyan	Red	Green	Absorption
Green	Yellow	Green	Blue	Emission
Green	Yellow	Blue	Green	Absorption

Table 1. The State Transition Table for the 3 Quarks of a Nucleon.

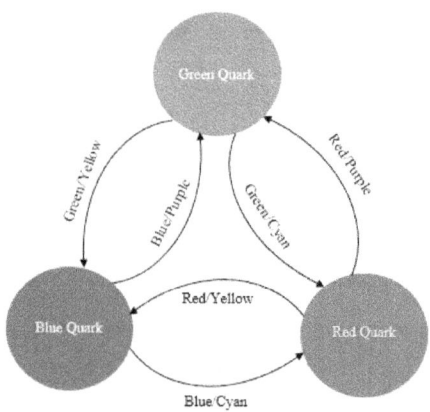

Figure 9. The Gluon Color Exchange Mechanism of the Quarks and Antiquarks.

The strong nuclear force is realized through the exchange of gluons. Every quark or antiquark may emit a gluon which may be absorbed by another quark or antiquark. The strong nuclear force shares this characteristic with electromagnetism with the emission and absorption of a photon, the carrier of the electromagnetic force, between two charged particles. Similarly the gluons are the carriers of color charges between quarks or between antiquarks. Thus, the gluons are the gauge bosons that mediate the strong nuclear force between quarks or between antiquarks.

§ 5. Is a nucleon a Mealy deterministic finite-state machine?

A finite-state machine, or a finite-state automaton, is a mathematical model of computation. It is a theoretical state machine that can be in precisely one of a finite number of states at any given time. The finite-state machine could transition from one state to another state in response to some inputs. A finite-state machine consists of a list of its states, an initial state or a start state, and the inputs that causes each transition. There are two kinds of finite-state machines, a deterministic finite-state machine and a non-deterministic finite-state machine. A deterministic finite-state machine can be designed to be like any non-deterministic finite-state machine. Furthermore, a deterministic finite-state machine can be designed with or without an output. Two important properties of a finite-state machine are that it is the simplest model of computation and it has a very limited memory.

In the computational field of theoretical computer science, a deterministic finite-state machine is a finite-state machine that can run through a state sequence specifically determined by a string of symbols, accepting or rejecting any symbols. So, the computation run is specific or unique.

Let us consider an illustration for a deterministic finite-state machine using a state diagram. In this example of the finite-state machine, there are three final states: Red, Green, and Blue, which are represented graphically by circles, for the color quarks of a nucleon. The finite-state machine takes a finite sequence of 0's and 1's as its input. At each state, there is a transition arrow leading out to the next state for either a "0" or a "1". Upon reading a symbol, a finite-state machine transitions deterministically from one state to another by following the corresponding transition arrow.

For example, a red quark may emit a red-antiblue gluon, turning the red quark blue, while turning the blue quark red, or a red-antigreen gluon, turning it green, while turning the green quark red, or a blue quark may emit a blue-antired gluon, turning it red, while turning the red quark blue, or a blue-anti-green gluon, turning it green, while turning the green quark blue, or a green quark may emit a green-antired gluon, turning it red, while turning the red quark green, or a green-anti-blue gluon, turning it blue, while turning the blue quark green.

In a Mealy machine, the output depends both upon the present state "AB" and the present input "I". Generally, it has fewer states than a Moore Machine. The value of the output function is a function of the transitions and the changes, when the input logic, or physical process in this case, on the present state is finished.

Mealy machines react faster to inputs. They generally react in the same clock cycle. In the following Mealy finite-state machine, the input "I" and the output "Y" are shown as "I/Y" next to each transitional arrow, where the color of every state of every quark in the output "Y" is designated with a value of "1" to represent a consistent potentially energized state.

Present State Q(t)	Next State Q(t+1)			
Quark Colors	Input = 0		Input = 1	
AB	State	Output	State	Output
Start 00	01	Green	10	Blue
Green 01	10	Blue	11	Red
Blue 10	11	Red	01	Green
Red 11	10	Blue	01	Green

Table 2. The State Table of a Mealy Machine for a Nucleon.

Let us define "Q" as the set of all states, Q = {00, 01, 10, 11}, "Σ" is equal to the inputs, Σ = {0, 1}, q_0 is the start state or initial state of creation of the nucleon, q_0 = {00}, "F" is the set of final states, F = {01, 10, 11}, and "δ" is the transition function from Q × Σ → Q. The transition states of "δ" are shown in the state table.

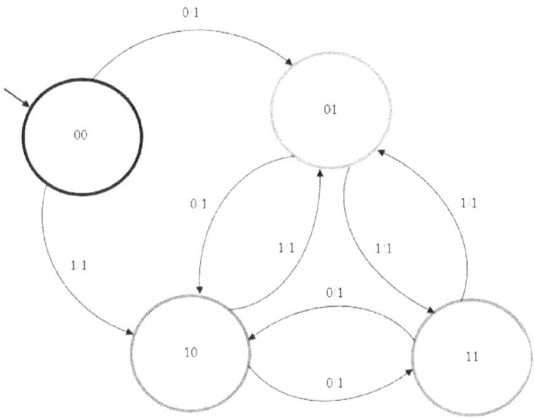

Figure 10. The State Diagram of a Nucleon as a Deterministic Finite-State Machine.

A Karnaugh map provides a pictorial method of grouping together expressions with common factors and eliminates unwanted variables. The Karnaugh map may be described as a special arrangement of a state table or truth table. The Karnaugh map provides a simple and straight-forward method of minimizing Boolean expressions.

A \ BI	00	01	11	10
0	1	1	1	1
1	1	1	1	1

Figure 11. The Karnaugh Map for the above State Diagram.

Thus, the Boolean function for the nucleon may be extracted from the Karnaugh map to obtain the following output:

$$Y = 1 \qquad (5.1)$$

Therefore, the Boolean function obtained is a tautology, a formula or assertion that is true in every possible interpretation. For example, every final state of every quark in a nucleon is red, green, or blue, or every final state of every quark in a nucleon is not red, green, or blue, is always true, regardless of the color of the final state of the quark. Hence, in this context, it seems that nature may also be deterministic at its fundamental level. In such case, one could say that "the creator of all there is" has retained the quantum right to play dice or not. Furthermore, a nucleon appears to be a consistent perpetual finite-state machine.

Last, but not least, it is interesting to mention that the above finite-state machine may have start states of "0", "1", or what is called a linear combination of "0" and "1", a superposition of two inputs for two quark color states, in our context. This fluid combination of amplitudes is at the core of quantum computers. Before one measures a qubit, or a quantum bit, it exists in a general state of superposition, a quantum version of a probability distribution, where each qubit has some amplitude relative probability for being zero, and some amplitude relative probability for being one. Superposition allows a quantum computer to store and manipulate a vast amount of data. Aside, n qubits = 2^n classical bits.

$$|\Psi\rangle = \alpha|0\rangle + \beta|1\rangle \qquad (5.2)$$

where α and β are probability amplitudes that can be, in general, complex numbers.

By Born's rule, if we measure a qubit in the standard basis, the probability of outcome $|0\rangle$ with a value of "0" is $|\alpha|^2$ and the probability of outcome $|1\rangle$ with a value of "1" is $|\beta|^2$. Because the absolute squares of the amplitudes represent probabilities, it follows that α and β must be constrained by the equation $|\alpha|^2 + |\beta|^2 = 1$.

The amplitude describes the amount of each quark color state in the qubit, and the phase describes the path that is being followed, since the phase is cyclical, the number of paths can be represented by a Bloch sphere.

Hence, a qubit can be in any quark color state represented by a point on a sphere, while a classical bit can only be either at the North pole or the South pole of the sphere. A classical bit only exists at either the North pole or South pole of the Bloch sphere, that is why a classical bit is a very strong and effective method of storing information.

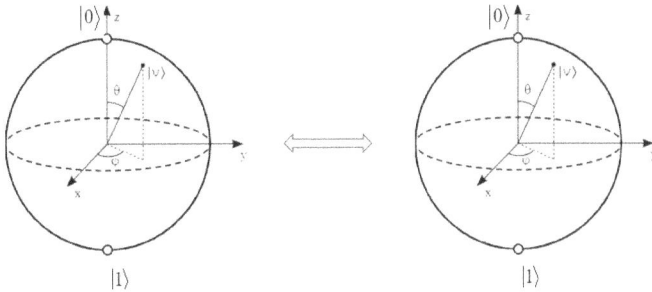

Figure 12. An Illustration of two Entangled Quantum Bits within the Boundary of a Nucleon.

Moreover, two or more qubits that are in a closed state of superposition can be entangled, endowing their final outcomes when measured to be mathematically related or correlated. Two qubits can provide four bits of information that specify the state of a specific quantum system. However, for a quantum computer to be useful, one needs to measure the information from the qubits, or in our case, from each nucleon, for 3^n values. A generalized qubit can have "N" states, or N^n values. So, when the quantum system is measured it collapses into the classical state of "0" or "1".

So, let us consider four possible states of these two entangled nucleons, to obtain

$$|\uparrow\uparrow\rangle + \frac{1}{\sqrt{2}}(|\uparrow\downarrow\rangle + |\downarrow\uparrow\rangle) + \frac{1}{\sqrt{2}}(|\uparrow\downarrow\rangle - |\downarrow\uparrow\rangle) + |\downarrow\downarrow\rangle \qquad (5.3)$$

The rotation, frequency, orientation, and count of the amplitude $|\Psi\rangle$ during an interval would be an indicator of the input value of a probable measurement for "0" and "1". The northern hemisphere of a nucleon represents input state $|0\rangle$ "ket zero" and the southern hemisphere of a nucleon represents input state $|1\rangle$ "ket one" after a measurement is performed. An amplitude indicator points from the center of a color quark triad within the nucleon toward a color state that exists at a specific instant of a measurement. It would be possible to have more than one amplitude indicator at a given instant.

It is possible to fully encode one bit in one qubit. Nonetheless, a qubit can hold more information, for example, up to two bits using superdense coding.

As a measurement is made to a quantum system using interference of waves, the amplitudes become probabilities. So, an observation, or a measurement, can extract an answer from a quantum system that is not a random outcome of probability.

The interference of the quarks in a nucleon could be harnessed by a deterministic sequence of qubit gates to cause the amplitudes to interfere constructively. This algorithmic effect elevates the probability of detecting one of the right answers.

A quantum computer does not have to perform faster than a classical computer in every possible case, but its number of operations to arrive at the result is exponentially small in comparison. So, its advantage is not the speed of an individual operation, but the total number of operations to arrive at the result in particular algorithms.

Could a nucleon ever become a quantum computer through some advanced technology?

The three color states of the quarks within the nucleon are the three probable states of the quantum bit. The rotation, frequency, and orientation of each color state of the color state triad within the boundary of the nucleon may be measured with interference to yield an outcome at a specific interval of the frequency.

Thus, each nucleon may yield three quantum qubits. Each entangled pair of nucleons may yield six qubits, and so on. For example, two qubits could be represented in a four-dimensional linear vector space spanned by the following product basis states:

$$|00\rangle = \begin{bmatrix} 1 \\ 0 \\ 0 \\ 0 \end{bmatrix} \quad |01\rangle = \begin{bmatrix} 0 \\ 1 \\ 0 \\ 0 \end{bmatrix} \quad |10\rangle = \begin{bmatrix} 0 \\ 0 \\ 1 \\ 0 \end{bmatrix} \quad |11\rangle = \begin{bmatrix} 0 \\ 0 \\ 0 \\ 1 \end{bmatrix} \quad (5.4)$$

$$\sum_{N=0}^{1}\left(|NN\rangle+|N\bar{N}\rangle\right)\equiv\begin{vmatrix}1&0&0&0\\0&1&0&0\\0&0&1&0\\0&0&0&1\end{vmatrix} \quad (5.5)$$

It would be a great leap forward in quantum computing if the nucleons could be measured in an atom, or in many atoms simultaneously. The logic combinations of the quantum gates would define the sequences of operations of the quantum computer. Nonetheless, the quantum computer may be constructed as a multi-dimensional quantum computer, or computing tesseract, for multi-dimensional processing of information both through space and time.

§ 6. So, what is the right number of gluons?

In color string theory, there are nine gluons that are possible because the photon is a gauge boson that also consists of gluons. Each gluon is a color-anticolor combination or pair. Color charges have three distinct colors and three distinct anticolors, and each possible combination represents a gluon pair.

A quark may emit a gluon, changing its own color in the process, and the emitted gluon may be absorbed by another quark that changes its color. There are three color pairs of color-anticolor of the same color that may combine in two triplets and seven other combinations of doublets of color pairs of color-anticolor of different color. Even though gluons are illustrated as springs, each gluon has a color charge.

There are three quarks in a nucleon that sum up to a colorless state, namely, red, blue, and green. Hence, in order to have a colorless state, to reiterate, the following exchanges would take place: a red quark may emit a red-antiblue gluon, turning the red quark blue, while turning the blue quark red, or a red-antigreen gluon, turning it green, while turning the green quark red, or a blue quark may emit a blue-antired gluon, turning it red, while turning the red quark blue, or a blue-anti-green gluon, turning it green, while turning the green quark blue, or a green quark may emit a green-antired gluon, turning it red, while turning the red quark green, or a green-anti-blue gluon, turning it blue, while turning the blue quark green. *Thus, the stable color-anticolor doublets are crucial for the color exchange mechanism of the quarks and antiquarks to maintain a colorless state.*

The above shows the exchange of six of the nine gluons. So, what happens to the rest of the gluons? The red-antired, blue-antiblue, and green-antigreen may form a photon or another color triplet.

§ 7. *The Color Exchange Mechanism.*

The color exchange mechanism would not exchange a color-anticolor pair, because the red quark that emits a red-antired pair would remain red without a quark that would absorb it since the other blue or green color-anticolor pair would not have an antiblue gluon or an antigreen gluon to cancel it to manifest an overall colorless state. In color string theory, the quantum gluons with the same quantum states may mix together. So, the three color-anticolor pairs of gluons may form triplets.

Color doublets or triplets exhibit the color-anticolor combination for a single color charge with a negative color-anticolor combination of a different color charge, or two single color charges with two negative color-anticolor combinations of a different color charge, to manifest a stable real gluon. If there were three color-anticolor combinations that are colorless, they would manifest a color triplet that is colorless as a single color state.

The following are examples of gluons of a color doublet or two color triplets:

$$\frac{|(red - antired) - (blue - antiblue)|}{\sqrt{2}} \quad (7.1)$$

$$\frac{|(red - antired) + (blue - antiblue) - 2 \cdot (green - antigreen)|}{\sqrt{6}} \quad (7.2)$$

$$\frac{|(red - antired) + (blue - antiblue) + (green - antigreen)|}{\sqrt{6}} \quad (7.3)$$

The actual color charges are arbitrary, but the color-anticolor combinations are mutually exclusive. The third color-anticolor combination is of the photon "γ^0" as predicted by color string theory. The photon is a color singlet state. From previous research, if the color state of a photon is measured, there would be equal probabilities of the state of the photon being red-antired, blue-antiblue, or green-antigreen. The photon is colorless, massless, may have relativistic mass, and may be virtual or non-physical, the possibility for the photon to be a color triplet is real. A photon is tricolor that is why is colorless. A tachyon may also be represented as a color triplet that is colorless with imaginary mass, and a negative sign due to intrinsic parity. (Nieves, 2020)

Hence, not all gluon matrices are traceless, which means that U(3) may be equal to SU(3) for spatial dimensions with the strong nuclear force governed by U(3). Then, there would be a second colorless gluon that behaves like a second photon. Would a tachyon fill the role of a second photon or would it be the i-partner of a relativistic photon?

A photon riding the retarded spatiotemporal wave and another riding the advanced spatiotemporal wave, a photon and its anti-photon (a tachyon). A photon and its copy have the phase difference of π and constitute a phase-entangled photon pair.

§ 8. The Particles and Anti-particles of the Current Standard Model.

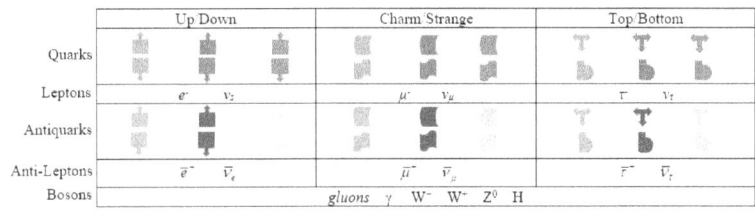

Figure 13. The Particles and Anti-Particles of the Current Standard Model.

Even though the current standard model agrees with most of the experimental results with a high degree of accuracy, it still has the following shortcomings: the current standard model does not predict gravitation, the current standard model does not unify the fundamental forces of nature, there is no particle predicted by the current standard model to explain the missing fermionic matter in our universe, the mass of the Higgs particle is unstable with respect to quantum corrections, and there are a few experimental results that do not agree with the predictions of the current standard model. Hence, "a Dynamic Theory of Space-Time" with the Gluon Standard Model may provide an ample theoretical framework to explain the shortcomings of the current standard model. (Nieves, 2020)

The strong nuclear force mediates the quarks and antiquarks with the combinations of the three colors and the three anticolors of the gluons. Six of the gluon pairs are uncomplicated, each has a color-anticolor combination that has a different anticolor for the corresponding color. The color-anticolor combinations of doublets are added together with a negative sign between the two color charge pairs. The triplet color combinations are colorless and may be complex; that is, real, imaginary, or both.

The Gluon Standard Model may be described by Group Theory, with the predictions of mathematical theory superbly described by the strong nuclear force. The properties of color charges are straightforward and fundamental to both electromagnetism and gravitation. The set of all possible combinations of gluon pairs endows every combination of quark or antiquark in our physical reality.

Electromagnetism, gravitation, and color string theory have a force of attraction and repulsion. Electromagnetism has positive and negative charges, the strong nuclear force has nine gluon combination to interact or not interact within every combination of quarks and antiquarks, and gravitation has an attractive gravitation force for celestial objects like a black hole or anti-gravitation from the corresponding hypothetical white hole. A planet's gravitation may be considered inward (attractive) and negative while a white hole may be considered to have anti-gravitation (repulsive) that is outward and positive.

§ 9. The Neutrino Problem of the Current Standard Model.

Why has progress in the foundation of physics stalled?

Is it time to move beyond the current standard model of physics?

The current standard model may be regarded as noticeably flawed, and it has some different shortcomings. The missing fermionic matter may not consist of real particles, so it may be necessary to change the current standard model to the gluon standard model and its related theories. The flaw of the current standard model may originate from our misunderstanding of gravitation. For example, the problem of "μ", the g-factor or $g-2$, is intriguing even though it has not reached five sigma, and its horrendous complicated calculation. The problem with the prediction of the current standard model motivates the search for new physics, but there are other problems with the current standard model. The current standard model does not include gravitation, so we know that we need to combine the current standard model with gravitation because the known particles do gravitate but the current standard model does not tell us how that works. Hence, it is possible to suggest that a quantum gravity theory is needed.

The neutrinos of the current standard model are peculiar. Neutrinos are somewhat hard to detect in particle accelerators, and are really perplexing since all the other particles that have masses of the current standard model have a left-chiral and a right-chiral version of themselves, but only the left-chiral versions of the neutrinos have been seen, and that is a problem, because we need both versions of laterality to endow particles with the property of mass. It is known that a neutrino has mass, so how does the mass of the neutrino come about? Consequently, there is something about neutrinos missing in the current standard model, either there is a right-chiral neutrino which is just so heavy that it has not been seen yet or there is something peculiar with neutrinos, and neutrinos are just different to all the other particles in the current standard model.

There is also an anomaly in the LSND and MiniBooNE experimental data, and related data from the Opera detector at CERN since 2008, the Oscillation Project with Emulsion-Tracking Apparatus, which looks for neutrino oscillations. The Opera experiment was aiming at demonstrating the tau neutrino appearance in a muon neutrino beam due to neutrino oscillations. That result was achieved with a significance larger than five sigma. The experiment is unique in its capability of detecting all three neutrino flavors. The experimental data revealed a signal for neutrino masses, that has reached six sigma, that cannot be explained with the current standard model. Thus, it is reasonable to say that this experimental result may indicate that the current standard model may need to be modified. It is possible to modify the current standard model by introducing the gluon standard model with additional particles while employing the same mathematical framework. The gluon standard model, like the current standard model, is based on a quantum field theory which includes quantum mechanics on the entire measurement process with a clear understanding of how a measurement is performed. The impact of the gluon standard model would usher in new methods of theory development to the path of the underlying new physics such as "A Dynamic Theory of Space-Time". (Nieves, 2020)

Why are there only a left-chiral version of the neutrino or a right-chiral version of the antineutrino?

Laterality refers to two properties of a particle: the direction of its spin and how it relates to the direction the particle is traveling. We do not observe the right-chiral neutrinos directly because, to good approximation, only left-chiral neutrinos interact with the weak force, and the weak force is the only mechanism we have observed neutrinos to interact with at all. Most particles come in two varieties: ones that spin clockwise and ones that spin anticlockwise. Neutrinos are the only particles that seem to just spin anticlockwise.

The relative orientations of spin and linear momentum for neutrinos and antineutrinos is apparently fixed and intrinsic to the particles. For neutrinos the spin is always opposite the linear momentum and this is referred to as a left-helicity spin, whereas the antineutrinos always have a right-helicity spin. And every neutrino we have ever observed moves at speeds indistinguishable from, or very near, the speed of light. The existence of right-chiral neutrinos is theoretically well-motivated, because the known active neutrinos are left-chiral and all other known fermions have been observed with both left and right chirality. In physics, the geometric property of a particle may have chirality, or may be chiral, if it cannot be mapped to its mirror image by any combination of rotations, translations, and some conformational changes. Right-chiral neutrinos may also explain in a natural way the small active neutrino masses inferred from neutrino oscillation. Neutrinos are one of the most abundant particles in the universe. Because neutrinos interact very weakly with matter, they are incredibly difficult to detect. They have only been observed to interact through the weak force, although it is assumed that they also interact gravitationally, while light interacts strongly with matter, especially with electrons. Neutrinos and electrons are leptons. Quarks and leptons, as well as most composite particles, like protons and neutrons, with an odd half-integer spin, are fermions. Neutrinos are very hard to detect because they have no electric charge. But when a neutrino passes through matter, if it hits something dead-on, it will create electrically charged particles. And those can be detected.

How is a neutrino its own antiparticle?

Neutrinos come in three flavors: electron neutrino, muon neutrino, and tau neutrino. They also have corresponding antiparticles, collectively named antineutrinos. Neutrinos however fall into a category called leptons. Leptons and quarks are fermions that constitute matter. Particles have mirror image particles called antiparticles, or antimatter. These antimatter particles have the same mass as the matter particles we know, but are opposite in every other way. The i-partner symmetry exhibits the reversal of all of the quantum properties, such as spin and charge. An antineutrino is the antiparticle partner of the neutrino, meaning that the antineutrino has the same mass but opposite charge of the neutrino. Although neutrinos are electromagnetically neutral, they have no electric charge and no magnetic moment, they may carry another kind of charge: a lepton number. In particle physics, the lepton number denotes which particles are leptons and which particles are not. The lepton number $\{0, +1, -1\}$, or the lepton charge, is a conserved quantum number in all particle interactions. Any neutrino would have lepton number of "+1", while its antineutrino would have a lepton number of "−1". However, it is possible that the lepton number is also not conserved in nature. In that case, there is nothing to distinguish a neutrino from its antineutrino, not the electric charge, not the lepton number, and not anything else. Moreover, an isolated neutrino does not turn into an electron. That would violate charge conservation, among other things.

The brilliant physicist Ettore Majorana in 1937 proposed the theory that neutrinos with mass might be able to turn into their antiparticles and back again. One particle that is its own antiparticle is the photon, a particle of light. Another is the neutral pion, which is made up of quark-antiquark pairs, and the gluon, which glues quarks together. Scientists would refer to such a neutrino, which is identical to its antineutrino, as a Majorana neutrino. Scientists are trying to solve this case by performing tricky experiments that require extremely cold, clean conditions. This research looks for a very rare predicted process called neutrinoless double beta decay that can occur only if neutrinos are Majorana particles.

Neutrinos were hypothesized in 1931 by the eminent physicist Wolfgang Pauli to resolve a crisis in physics that threatened the bedrock principle of the conservation of energy. Wolfgang Pauli hypothesized that the nucleus emitted a second particle that could carry away this unaccounted-for energy. A nucleus undergoing beta decay emits a neutrino with the electron. Neutrinos are created by various radioactive decays; the following list is not exhaustive, but includes some of those processes: beta decay of atomic nuclei or hadrons, natural nuclear reactions such as those that take place in the core of a star, or when cosmic rays or accelerated particle beams strike atoms. Pauli's neutrino is now identified as the electron neutrino, while the second neutrino is named the muon neutrino, and the third neutrino is called the tao neutrino.

Neutrinos mix, and also quarks, that is, they turn into each other back and forth, as they propagate through the spatiotemporal medium. It is theorized that the combination of neutrino or antineutrinos particle-waves during oscillations may result in the superposition of the three unobserved categories of neutrinos or antineutrinos that produces a resultant flavor of a particle-wave that may endow the transmutation of neutrinos or antineutrinos (electron↔muon, muon↔tao, tao↔electron) as they propagate as tardyons through the spatiotemporal medium. The superposition of three unobserved categories of neutrinos or antineutrinos (primo, secondo, terzo) may explain the transmutation mechanism of the three observable flavors. Each unobserved category has its own characteristic mass, angular frequency, velocity, and phase angle, which after particle-wave superposition and phase angle shifting during oscillations would result in transmutations between the three observable neutrino flavors. Each observable neutrino flavor may be expressed as:

$v_{flavor} = a_1 v_1 + a_2 v_2 + a_3 v_3$. The complex coefficient "a_χ" of each unobserved category of neutrino, or antineutrino, in the superposition includes a relativistic effect, $m_\chi / \sqrt{1-(v/c)^2}$.

What is in the mirror?

Nature is symmetrical at the fundamental level. Symmetry rules the characteristics of the interactions of particles, and impedes the existence of massive particles. Physical reality has hidden symmetry where quarks, leptons, weak interaction bosons, and the Higgs boson, but not gluons or photons, acquire mass. There also color supersymmetry that is theorized that can stabilize the mass of the Higgs boson, which may be regarded as a general symmetry that may be imposed on a quantum field theory such as "A Dynamic Theory of Space-Time." (Nieves, 2020)

Color Supersymmetry is a coupling theory, the closer the masses of i-partners to particles, the greater the agreement with the gluon standard model. The mass of the i-partner of the Higgs boson does not have to be so much greater and decoupled. The experiments from the 1950s demonstrated that fundamental particles do not behave the same way when their spatiotemporal orientations are transposed into their mirror images. All particles were expected to obey the mirror or parity symmetry, breaking the mirror symmetry down. So, it was proposed that the symmetry might be completed by unobservable particles as part of an unknown bigger picture. Color Supersymmetry (COSUSY) provides i-partners that fulfill this early expectation. The mirror symmetry was expected to exist in our physical reality. COSUSY is a six-dimensional theory with a (3 + 3) formalism. The i-partners are reflections that may be regarded as imaginary partners that mirrors the matter in our universe, unobserved but sensed by particles. From astronomical research in the 20th century, the expectation of unobserved matter, is equivalent to a missing fermionic matter that has moved whole galaxies by its gravitational influence. The approximate distribution of the unobserved matter or missing fermionic matter was mapped out and estimated to be five times that of the visible fermionic matter. But researchers have not found out the missing fermionic matter in the early part of the 21st century yet.

The problems of the unobservable missing fermionic matter

Observations have accumulated from the beginning of the twenty first century that the unobservable missing fermionic matter cannot explain. The unobservable missing fermionic matter is no longer the simplest parametric explanation. These observations beg for questions of science and their explanations without fiction. The behavior of particles and fields changes from the scales of galaxies, to clusters, to filaments, and to the early universe. A theory needs to include a kind of phase transition that explains why and under which circumstances the behavior of these additional particles, or fields, changes so we need to express this behavior in two distinct sets of equations for the unobservable missing fermionic matter and the unobservable missing field energy, or in a unique distinct set of equations like the six-dimensional EFEs. A theory that deals with condensed matter physics, the physics of solids, gases, and fluids. A theory that combines the attributes of the unobservable missing fermionic matter and those of modified gravity.

Some of the problems of the unobservable missing fermionic matter are, but not limited to, the following: the unobservable missing fermionic matter predicts too many small galaxies, these are small-scale galaxies that orbit around a larger host, known as satellite galaxies. For example, the Milky Way only has a few dozens, but it should have hundreds, the small satellite galaxies are often aligned in planes. The unobservable missing fermionic matter does not explain why. It is also known from observation that the mass of a galaxy is correlated from three and a half to the fourth power of the rotation velocity of the outermost stars. This is called the baryonic Tully–Fisher relation and it is just an observational fact. The baryonic Tully–Fisher relation is a factual relationship between the baryonic mass, the sum of its mass in stars and gas, or the intrinsic luminosity of a spiral galaxy and its asymptotic rotation velocity or emission line width. The unobservable missing fermionic matter does not explain it. The unobservable missing fermionic particle matter predicts a density in the cores of small galaxies that peaks, whereas observations say the distribution should be flat. If you look at the rotation curve of a galaxy, then for every feature in the curve for the visible emission, like a bump or a wiggle, there is also a feature in the rotation curve. This is known as Renzo's Rule. Again, hat is an observational fact, but it would be something inexplicable to think that most of the matter in galaxies is the unobservable missing fermionic matter. The unobservable missing fermionic matter should remove any correlation between the rotation curves and the luminosity. Then, there are collisions of galaxy clusters at high velocity, like the "El Gordo" cluster or the bullet cluster. These are difficult to explain with the unobservable missing fermionic particle matter, because the unobservable missing fermionic matter creates friction and that makes such high relative velocities extremely unlikely. The bullet cluster is a problem for the unobservable missing fermionic matter, but not evidence for it. In the book "A Dynamic Theory of Space-Time", there is a possible explanation to the missing fermionic matter in our universe as a correction to the Einstein Field Equations for six-dimensional space-time.

9.1 The Peculiar Neutrino of the Gluon Standard Model.

The neutrino is considered a lepton with mass with a half-spin. Let us theorize that the neutrino has positive intrinsic parity with a neutral electrical charge, and the antineutrino has a negative intrinsic parity. The intrinsic parity is a phase factor that arises as an eigenvalue of the parity operation, which is a reflection, or a mirror symmetry, about the origin. The neutrino is theorized to be left-chiral or right-chiral. Let us denote the two polarization states of the neutrino to be positive or negative. The neutrino has three flavors, each flavor is either ± when the charge of the conjugate antineutrino is ∓. Thus, any single pair, of a neutrino and its conjugate antineutrino, may have positive or negative polarity according to the polarization correspondence principle. The helicity (polarity) of a neutrino describes a combination of the spin and the instantaneous linear motion that is Lorentz invariant, i.e. the helicity has a value that is the same in all inertial reference frames. If the spin vector of a neutrino points in the same direction as the momentum vector, the helicity (polarity) is positive (right-helicity), if the spin and momentum vectors point in opposite directions, the helicity (polarity) is negative (left-helicity). *The helicity of a neutrino, a massive particle, unlike a photon, is not equal to its quantum mechanical chirality. A neutrino is chiral if it is indistinguishable from its reflection in a plane mirror. So, chirality is built into the neutrino, but helicity is a matter of perspective.*

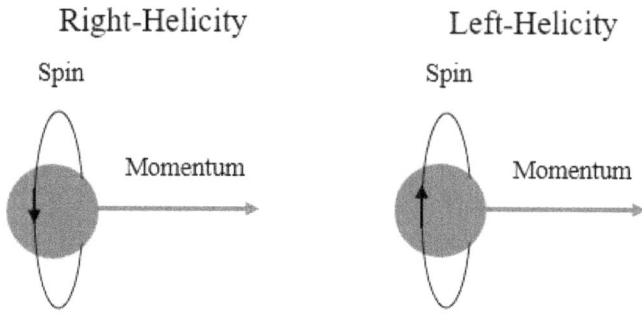

Figure 14. An Illustration of Helicity (Polarity) for a Particle.

The following figure of a circle represents the complex phase of the quantum state of a particle as a particle rotates, and the value of the phase angle changes around the circle. As a particle rotates 360 degrees, the particle moves only halfway around the circle in a sense of rotation that depends on the chirality of the particle. If a left-chiral or a right-chiral particle rotates 360 degrees, both would reach a phase value of −1 at halfway around the circle on the complex plane. The left-chiral particle would move in a clockwise direction while the right-chiral particle would move in a counterclockwise direction.

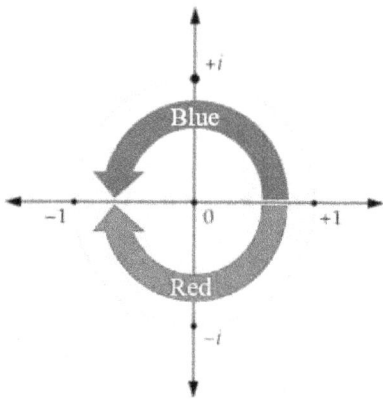

Figure 15. An Illustration of the Right Chirality (Blue Arrow) and the Left Chirality (Red Arrow) of a Particle.

The phase angle of the wavefunction of a particle may be shifted when a particle rotates in a manner that depends on the chirality of the particle.

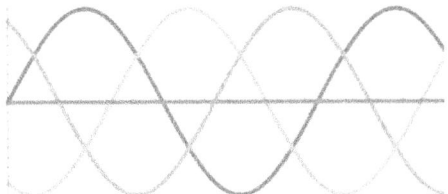

Figure 16. The Effect of Chirality on The Quantum Mechanical Phase Shift of the Wavefunction of a Particle (center green arrow). Left (blue) arrow is the rotation effect on a left-chiral fermion, and right (red) arrow is the rotation effect on a right-chiral fermion.

The quantum wavefunction of a fermion is shifted when a fermion rotates causing a polarity quantum event, when left-chiral and right-chiral fermion are shifted in opposite directions. A particle's inherent quantum properties are related in a profound manner. When the weak interaction endows mass to fermions, the effect of chirality becomes more significant.

The neutrino cannot be superposed onto its reflected image; so, the neutrino is achiral. Human hands are chiral. The combination of spin and instantaneous linear motion (momentum) of the neutrino describes its chirality (laterality) or helicity (polarity). For the massive neutrino with neutral electrical charge, the helicity (polarity) is not the same as the chirality (laterality). Chirality or helicity of the massive neutrino are either positive (right-handed, counterclockwise) or negative (left-handed, clockwise) corresponding to the polarity based on the magnitude of the Planck quantum charge, q_P, carried by a single neutrino or electron. By rotational convention, if one imagines a massive round wall clock that is tossed like a frisbee to the right, with its spin vector defined by its hands and its dial facing up, such a clock would have negative chirality and helicity. Opposite massive neutrinos have opposite chirality and helicity. Opposite massive neutrinos attract, equal massive neutrinos repel. Opposite chirality and helicity complement, equal chirality and helicity separate. The relationship of the opposite massive neutrino, or that of chirality and helicity between massive neutrinos, in a spatiotemporal realm of scales where electrons emerge from the weak interaction. It is no coincidence that the electromagnetic force and the weak interaction are two aspects of a single electroweak force. (Nieves, 2020)

Chirality and helicity designate the laterality and polarity of a neutral electrical charge, such as a massive neutrino. *The weak interaction couples to the left-chiral fermions, whatever effect this has on helicity depends on the kinematics of the interaction.* Charge itself is neutral. Charge is spatiotemporal in nature. A Coulomb of charge consists of a unit of length times its orthogonal conjugate unit of time. Charge is an emerging two-dimensional spatiotemporal manifestation. The Planck quantum charge is applicable at the scale of a neutrino. Any charge on the neutrino is a multiple of the Plank quantum charge, $t_p l_p$.

It is hypothesized that chirality produces a laterality, and spin produces spatiotemporal frame-dragging, about a neutrino that is opposite for positive or negative, providing attraction or repulsion. Opposite neutrino charges provide complementary spatiotemporal frame-dragging, decreasing the distance between the charges. Like neutrino charges provide repulsive spatiotemporal frame-dragging, increasing the distance between the charges. The laterality of chirality allows superposition of the neutrino charges according to the Gluon Standard Model.

9.2 The Transmutation of Fermions by the Electroweak Force.

The important function of the weak nuclear interaction is to cause particle transformations, more specifically, it turns particles like quarks into particles like leptons. However, it has been very difficult to point to exactly what and how the weak nuclear interaction does what it does. the weak interaction plays a fundamental role in the universe. The odd-one-out parity status of the weak interaction provides researchers with a valid reason into the pursuit of new physics, beyond the current standard model. This pursuit involves research into the weak force and into particles such as neutrinos and gluons, and gravitation.

During a beta-plus decay of a proton into a neutron, the transformation of a up quark to a down quark occurs, while emitting a W^+ gauge boson, with an electrical charge of $+1e$, drawing energy and color charges from the positron to its gluonic fields, to render an electron neutrino.

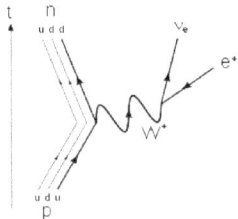

Figure 17. Feynman Diagram of Electron Neutrino Emission.

Similarly, during a beta-minus decay of a neutron into a proton, the transformation of a down quark to an up quark occurs, while emitting a W⁻ gauge boson, with an electrical charge of $-1e$, transferring energy and color charges from its gluonic fields, to transmute the electron anti-neutrino (a lepton) into an electron.

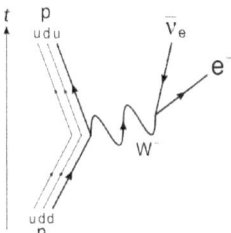

Figure 18. Feynman Diagram of Electron Emission.

The $W^{\pm}$ gauge bosons are gluonic mechanisms to draw or transfer energy from gluonic fields, and to transfer color charges, during lepton transmutations. The gauge bosons and their symmetries provide gauge transformations to quantum particles that leave the quantum properties and field interactions of the particles unchanged in our universe. When the same phase shift, a global phase shift, is provided across the whole wavefunction of a particle, to the real and imaginary components of the wavefunction, all the observables are unchanged. Global phase is a gauge symmetry of the particle or system. The square of the wavefunction determines the position of a particle, even though the phase shift itself may be unobservable.

From previous research, it is possible to suggest that when a photon affects an electron in an atom, whereby the electron may absorb some of the energy in the incident photon to change its orbit or be ejected from its atom, the photoelectric effect may be causing the onset of the quantum process of gluon parity at the atomic level. An electron may also be manifested in an atom through the doubling of gluon pairs increasing the overall color charge of the resulting electron. In particle accelerators, electrons and positrons are produced through the process of pair production. In this process a high-energy photon interacting with the electroweak field of a heavy charge creates an electron and a positron.

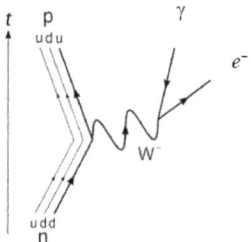

Figure 19. Feynman Diagram of Electron Emission by a Photon.

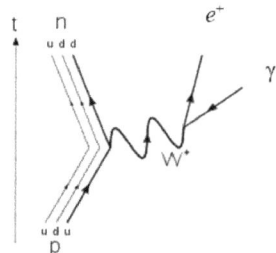

Figure 20. Feynman Diagram of Positron Emission by a Photon.

A photon can spontaneously degenerate into a particle with mass and its antiparticle in a process known as pair production. In this process, the energy of the photon is completely transformed into the mass of the two particles. For pair production to occur, the discrete quantity of electromagnetic energy of a photon, must be at least equivalent to the mass of two electrons. When pair production occurs, photon energy in excess of this amount is converted into motion of the electron-positron pair.

Consequently, it is possible to theorize that as the symmetry of each gluon pair in a photon breaks during parity, the gluon field of each pair reverses itself, traveling in the opposite direction as an electron, as the primary carrier of electricity in solid conductors, according to the Gluon Standard Model. Thus, the electron travels in the direction of the positive charges of the field potential, but the photon travels in the opposite direction, that is, in the same direction of the electromagnetic field.

$$Photon + W^{\pm} \rightarrow Positron \ and \ Electron$$

$$r\bar{r} + b\bar{b} + g\bar{g} + W^{\pm} \rightarrow \bar{r}r + \bar{b}b + \bar{g}g \ \ and \ -r\bar{r} - b\bar{b} - g\bar{g} \quad (9.1)$$

Therefore, there is also the transmutation of the electron antineutrino into the electron, and the positron into the electron neutrino. Neutrinos and antineutrinos are produced in beta-decays to obey the rule of lepton conservation. The production of a charged lepton is always accompanied by the corresponding flavor of neutrino. In all weak interactions the following conservation rules are followed: the electric charge is conserved, the number of leptons minus the number of antileptons is conserved, the number of quarks minus the number of antiquarks is conserved, and the change of the flavor of leptons or quarks is allowed.

Hence, it is reasonable to theorize that the gluon pairs in the substructure of each flavor of neutrino, or charged lepton, is conserved. Consequently, let us formulate the gluonic equations for the electron neutrino and the electron antineutrino using their lepton numbers.

$$W^{-} \times Electron \ Antineutrino \rightarrow Electron$$

$$W^{-} \times (-1)(\bar{r}r + \bar{b}b + \bar{g}g) \rightarrow (+1)(-r\bar{r} - b\bar{b} - g\bar{g}) \quad (9.2)$$

$$W^{-} \times (-\bar{r}r - \bar{b}b - \bar{g}g) \rightarrow -r\bar{r} - b\bar{b} - g\bar{g} \quad (9.3)$$

Therefore, the W^{-} may be defined as the negative left-chiral weak interaction operator as follows:

$$W^- \equiv -i\hbar \frac{\partial}{\partial t} e^{-i\omega_{w-} t} \tag{9.4}$$

$$W^- \equiv \frac{\hbar^2}{2m_{w-}} \nabla^2 - V(\vec{r},t) \tag{9.5}$$

Applying the W^- operator on the electron antineutrino, we have

$$W^- \cdot \overline{v}_e = -i\hbar \frac{\partial}{\partial t} e^{-i\omega_{w-} t} \cdot e^{-i\omega_{\overline{v}_e} t} = -i\hbar \frac{\partial}{\partial t} e^{i(-\omega_{w-} - \omega_{\overline{v}_e})t} \tag{9.6}$$

$$W^- \cdot \overline{v}_e = -i\hbar \frac{\partial}{\partial t} e^{-i(\omega_{w-} + \omega_{\overline{v}_e})t} = i^2 \hbar \left(-\omega_{w-} - \omega_{\overline{v}_e}\right) e^{-i(\omega_{w-} + \omega_{\overline{v}_e})t} = -\hbar \omega_e e^{-i\omega_e t} \tag{9.7}$$

$$W^- \cdot \overline{v}_e = -\hbar \omega_e e^{-i\omega_e t} \tag{9.8}$$

Where the weak interaction operator W^- is an energy and waveform operator,

$$W^- \equiv -i\hbar \frac{\partial}{\partial t} e^{-i\omega_{w-} t} = -\hbar \omega_{w-} \frac{\partial}{\partial t} \left(\text{Cos}(-\omega_{w-} t) - i\text{Sin}(-\omega_{w-} t) \right) \tag{9.9}$$

$$W^- = -\hbar \omega_{w-} \left\{ -\text{Sin}(\omega_{w-} t) - i\text{Cos}(-\omega_{w-} t) \right\} = \hbar \omega_{w-} \text{Sin}\,\omega_{w-} t + i\hbar \omega_{w-} \text{Cos}\,\omega_{w-} t \tag{9.10}$$

$$W^- = \hbar \omega_{w-} \text{Sin}\,\omega_{w-} t + i\hbar \omega_{w-} \text{Cos}\,\omega_{w-} t = -i\hbar \omega_{w-} e^{-i\omega_{w-} t} \tag{9.11}$$

Similarly, the W^+ may be defined as the positive left-chiral weak interaction operator as follows:

$$W^+ \times \text{Positron} \rightarrow \text{Electron Neutrino}$$

$$W^+ \times (-1)\left(-\overline{r}r - \overline{b}b - \overline{g}g\right) \rightarrow (+1)\left(r\overline{r} + b\overline{b} + g\overline{g}\right) \tag{9.12}$$

$$W^+ \times \left(\overline{r}r + \overline{b}b + \overline{g}g\right) \rightarrow r\overline{r} + b\overline{b} + g\overline{g} \tag{9.13}$$

$$W^+ \equiv +i\hbar \frac{\partial}{\partial t} e^{-i\omega_{w+}t} \qquad (9.14)$$

$$W^+ \equiv -\frac{\hbar^2}{2m_{w-}} \nabla^2 + V(\vec{r},t) \qquad (9.15)$$

Applying the W^+ operator on the positron, we have

$$W^+ \cdot e^+ = +i\hbar \frac{\partial}{\partial t} e^{-i\omega_{w+}t} \cdot e^{-i\omega_{e+}t} = +i\hbar \frac{\partial}{\partial t} e^{i(-\omega_{w+}-\omega_{e+})t} \qquad (9.16)$$

$$W^+ \cdot e^+ = +i\hbar \frac{\partial}{\partial t} e^{i(-\omega_{w+}-\omega_{e+})t} = i^2 \hbar \left(-\omega_{w+}-\omega_{e+}\right) e^{i(-\omega_{w+}-\omega_{e+})t} = +\hbar \omega_{v_e} e^{-i\omega_{v_e}t} \qquad (9.17)$$

$$W^+ \cdot e^+ = +\hbar \omega_{v_e} e^{-i\omega_{v_e}t} \qquad (9.18)$$

Where the weak interaction operator W^+ is an energy and waveform operator,

$$W^+ \equiv +i\hbar \frac{\partial}{\partial t} e^{-i\omega_{w+}t} = +\hbar \omega_{w+} \frac{\partial}{\partial t} \left(\text{Cos}(-\omega_{w+}t) - i\text{Sin}(-\omega_{w+}t)\right) \qquad (9.19)$$

$$W^+ = \hbar \omega_{w+} \frac{\partial}{\partial t} \left(\text{Cos}(\omega_{w+}t) - i\text{Sin}(\omega_{w+}t)\right) \qquad (9.20)$$

$$W^+ = \hbar \omega_{w+} \left\{-\text{Sin}(\omega_{w+}t) - i\text{Cos}(\omega_{w+}t)\right\} = -\hbar \omega_{w+} \text{Sin}\,\omega_{w+}t - i\hbar \omega_{w+} \text{Cos}\,\omega_{w+}t \qquad (9.21)$$

$$W^+ = -\hbar \omega_{w+} \text{Sin}\,\omega_{w+}t - i\hbar \omega_{w+} \text{Cos}\,\omega_{w+}t = +i\hbar \omega_{w+} e^{-i\omega_{w+}t} \qquad (9.22)$$

It is interesting to note that the W^- operator shifts the phase angle of the interacting left-chiral electron antineutrino by $-90°$ to the left, and the W^+ operator shifts the phase angle of the interacting left-chiral positron by $+90°$ to the right. Thus, it is hypothesized that the modulation of the angular frequency "ω" of the $W\pm$ operators endows the resulting electron or electron neutrino with greater or lesser mass.

It is interesting that the Schrodinger equation may be expressed in terms of the W± boson as $-W^-[\Psi(r, t)] = i\hbar\omega_{w-}\ e^{-i\omega_{w-}t}[\Psi(r, t)]$, or $W^+[\Psi(r, t)] = i\hbar\omega_{w+}\ e^{-i\omega_{w+}t}[\Psi(r, t)]$, which describes the evolution of the wave function as the mathematical object that contains all the information of a particular particle or physical system.

Aside, the weak interaction W boson is a massive spin-1 and left-chiral particle, that assumes any of the three helicities: one is longitudinal W^0, and two are transverse states, W^- has left-helicity, and W^+ has right-helicity. Left-helicity or left-chiral is negative, and right-helicity or right-chiral is positive. When a W boson or other gauge boson acquires mass through the Higgs mechanism, this boson must also acquire a longitudinal polarization state that does not exist for a massless gauge boson. For a highly boosted W boson, there is a clear distinction between the transverse and longitudinal polarization states. However, how other elementary particles acquire mass from the Higgs boson is to couple to the Higgs boson, so those particles need to be left-chiral and right-chiral, and the Higgs boson needs to couple to both left-chiral and right-chiral particles together. That works for all particles except the neutrinos.

The neutrinos mix into each other as they propagate, so they must have masses. How do neutrinos acquire their masses? It is currently thought that either the neutrino is very heavy and requires a lot of energy to be created or neutrinos are Majorana particles (the left-chiral and right-chiral neutrinos versions are the same). The global fit of oscillation parameters determine the masses of the neutrino mixing, the wavelength of the mixing and the mixing angles to determine how much the neutrinos mix. Is the above hypothesis supported that the modulation of the angular frequency "ω" of the W± operators endow neutrinos with their masses? Consequently, more long-term experiments on neutrinos may be needed to arrive at a significant level of confidence.

What is the mass of an electron neutrino or an electron antineutrino that interacts with the W± boson?

$$\textit{Electron Neutrino or Electron Antineutrino Mass} \equiv \frac{10^{-54}\,Kg}{\sqrt{1-\frac{v^2}{c^2}}} \quad (9.23)$$

Where $v \sim c$ for a neutrino or an antineutrino, and the photon mass is about 10^{-54} Kg. The positron interacting with the W^+ boson is left-chiral, with a lepton number of -1, and a charge $+1$, and the electron neutrino produced is left-chiral, with a lepton number of $+1$, and a charge of 0. The Electron Antineutrino interacting with the W^- boson is left-chiral, with a lepton number of -1, and a charge of 0, and the Electron produced is left-chiral, with a lepton number of $+1$, and a charge -1.

Aside, the $W^{\pm}$ Boson mass is $\sim 1.433028377 \times 10^{-25}$ kg (~ 80.4 GeV), the electron mass is $\sim 9.10938356 \times 10^{-31}$ Kg, the electron neutrino is ~ 0.07 eV or $\sim 1.25 \times 10^{-37}$ Kg when $v \sim c$, the average neutrino mass is < 0.120 eV ($< 2.14 \times 10^{-37}$ Kg), with a 95% confidence level, for the sum of three neutrino flavors. The electron antineutrino has the same mass but opposite charge of the electron neutrino. The effective range of the weak interaction which is around 10^{-16} m to 10^{-17} m. At 10^{-18} m, the weak interaction has a strength of a similar magnitude, to the electromagnetic force. The Z^0 Boson mass is $\sim 1.625555857 \times 10^{-25}$ kg (~ 91.2 GeV), about 12.5% heavier the $W^{\pm}$ boson.

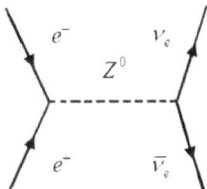

Figure 21. An electron–positron pair undergo annihilation, creating a Z^0 boson which subsequently decays into an electron neutrino and an electron antineutrino pair. The Z^0 boson decays in about twenty percent of the cases into a neutrino-antineutrino pair.

The neutrinos are unobservable at the present time and can only be measured when there is some energy or transverse momentum missing after the collision, since it is known that both energy and transverse momentum should be conserved in the collision.

The Z^0 boson is its own antiparticle. Thus, all of its flavor quantum numbers and charges are zero. The exchange of a Z^0 boson between particles, known as a neutral current interaction, leaves the interacting particles unaffected, except for a transfer of spin and/or momentum. Unlike for the weak interaction $W^{\pm}$ boson, the Z^0 Boson couples to both left-chiral and right-chiral states with different strengths, not equally.

So, where are all the right-chiral neutrinos and the left-chiral antineutrinos?

The weak interaction, or weak force, in terms of the electro-weak theory, as the mechanism of interaction between sub-atomic particles that causes radioactive decay and plays an essential role in nuclear fission. The weak interaction is sometimes referred to as flavordynamics. Moreover, the weak interaction is the only fundamental interaction that breaks charge parity symmetry. For instance, the weak interaction occurs within the boundary of a proton as a fundamental interaction of nature. (Nieves, 2020)

A neutron may beta-decay into a proton, an electron, and an electron anti-neutrino, through the weak interaction. Carbon-14 (6 protons and 8 neutrons) is an example of a beta-minus decay to Nitrogen-14, 7 protons and 7 neutrons, plus an anti-neutrino and an electron. Colors and anticolors play an essential role in the weak interaction at the deepest levels of nature. All weak interactions are very near-field gluonic interactions within the boundary of a particle. The exchange of $W^{\pm}$ and Z^0 gauge bosons not only causes the transmutation of a quark, i.e. changing a quark flavor, inside hadrons, but also changes the hadrons themselves. For example, a proton can decay into a neutron, transforming an up quark (udu), with an electrical charge of $+(2/3)e$, into a down quark (udd), with electrical charge of $-e/3$, while emitting a W^+ gauge boson, to transmute a positron into an electron neutrino. Another example is the electron capture, a common variant of radioactive decay, where a proton and an electron within an atom interact, the proton is changed to a neutron, an up quark is changed to a down quark, and the electron is changed to an electron neutrino and emitted. The $W^{\pm}$ and Z^0 bosons, together with the photon, are the gauge bosons of the electroweak interaction.

Therefore, it possible to make the following observations:

1. The weak interaction $W^{\pm}$ boson will couple to both left-helicity and right-helicity particles, such as an electron and a positron. Nonetheless, The $W^{\pm}$ boson will only couple to left-chiral electrons and left-chiral positrons and will not couple to right-chiral electrons or to right-chiral positrons.
2. During the weak interaction of the W^- and the electron antineutrino that renders the electron, there is symmetry breaking that transmutes the gluon pairs, breaking the parity symmetry of the gluon pairs to the transposed gluon pairs of the electron. Thus, when a left-chiral electron antineutrino interacts with the W^-, a similar symmetry breaking would occur to its gluon pairs to render a left-chiral electron. This reaction can happen in a neutron within an atom or a free-floating neutron.
3. During the weak interaction of the W^+ and the positron that renders the electron neutrino, there is symmetry breaking that transmutes the gluon pairs, breaking the parity symmetry of the gluon pairs to the transposed gluon pairs of the positron. Thus, when a left-chiral positron interacts with the W^+, a similar symmetry breaking would occur to its gluon pairs to render a left-chiral electron neutrino.
4. The electron neutrino and the electron antineutrino have the same chirality, and the electron and the positron have the same chirality, but they have opposite helicity (polarity), during the weak interactions in the cases under consideration, according to the polarization and chirality correspondence principles.
5. The weak interaction is the only interaction that violates P or parity-symmetry and the CP or the charge–parity symmetry due to the symmetry breaking that can transmute the gluon pairs. The unique weak interaction can also transmute the flavors of quarks, changing one type of quark into another. *This is the principle of the transmutation of charge and parity of the weak interaction.*
6. The coupling constant of the weak interaction is an indicator of weak interaction strength, about 10^{-6} to 10^{-7}, that is the ratio of the masses of the particles rendered by the weak interaction, $W^{\pm}$ boson. In the cases under consideration, the mass ratio of the electron antineutrino to the electron, or the mass ratio of the electron neutrino to the positron.
7. The particle pair production process converts radiant energy to matter, to conserve energy and momentum. As a photon interacts with a $W^{\pm}$ boson, the weak interaction renders an electron–

positron pair near a nucleus. Consequently, the electron could be left-chiral, like the $W^{\pm}$ boson, by the southpaw correspondence principle of the weak interaction, and its conjugate positron could be right-chiral according to the laws of energy and momentum conservation.

8. Particles can in principle be left-chiral or right-chiral. An experiment at the LHCb provided new evidence that the $W^{\pm}$ bosons that mediate the weak force are all left-chiral, or southpaw, with negative helicity. They interacted only with left-chiral quarks. That may provide a possible explanation why the $W^{\pm}$ gauge boson would transmute a photon into an electron or positron with charge and parity symmetry. Photons are massless particles with their chirality equal to their helicity, that is relativistic invariance, a photon appears to spin in the same direction along its axis of motion regardless of the point of view of the observer.

9. It is reasonable to theorize that all of the observable electron neutrinos produced by the weak interaction are the left-chiral electron neutrinos, may be due to the southpaw correspondence principle of the weak interaction $W^{\pm}$ boson. The weak interaction mechanism may not be producing right-chiral electron neutrinos, which may be very hard to detect, that may or may not exist in our universe.

10. The weak interaction mechanism in every atom utilizes left-chiral electron antineutrinos to produce electrons which are very common. The more electron antineutrinos that are utilized by the W^- boson, the more common would be to find the unutilized conjugate right-chiral electron antineutrinos. Electrons are found in every atom of matter that exists in the universe.

It is interesting to note that since the weak interaction mechanism does not interact with the right-chiral positron, or may not utilize the right-chiral electron antineutrinos, the preference for left-chiral particles or left-chiral anti-particles over right-chiral antiparticles of the weak interaction mechanism, may contribute to some extent to the preference and formation of matter over anti-matter in our observable universe.

§ 10. The Gluon Theory as the Substructure of Elementary Particles.

Gluons are massless like photons. Gluons are considered quantum fields. Quarks are made of massless Gluons, but they have mass! Nucleons are made of Quarks. The atom has mass. As a result, they can approach, but never reach, the speed of light in a vacuum. The Higgs mechanism is essential to explain the generation mechanism of the property of "mass" for gauge bosons (γ, g, $W^\pm$, Z^0). The photon "γ" is a gauge boson. The Higgs field gives mass to other fundamental particles such as electrons and quarks.

The strong nuclear force is the most potent in nature, trillions of times stronger than gravity. If energy is used to split a quark pair, new quarks are produced, this is how matter was produced when the universe formed.

Mass (GeV)	Fermions			Bosons
	First Generation	Second Generation	Third Generation	
10^3				
10^2			Top	$W^\pm$, ↕ H, Z^0
10^1				
10^0		Charm	Bottom, Tao	
10^{-1}		Strange, Muon		
10^{-2}	Up, Down			
10^{-3}	e			
10^{-4}				
10^{-5}	—	—	—	—
10^{-6}	—	—	—	—
10^{-7}	—	—	—	—
10^{-8}	—	—	—	—
10^{-9}			↕ν_τ	
10^{-10}	↕ν_e	↕ν_μ		γ
10^{-11}				g

Table 3. The Masses of the Particles of the Gluon Standard Model.

The photon consist of three gluon~anti-gluon pairs, with zero charge, $0e$, within a certain spatiotemporal volume or outer surface of the volume. Why would an electron not consist of three quarks (down, strange, bottom) with a charge of $3 \cdot (-e/3) = -e$, the 6 negative gluon~anti-gluon pairs that make up a photon or 3 quarks, within a certain spatiotemporal volume or outer surface?

The present expressions of electromagnetic fields of a photon properly describe the particle characteristics of a photon. The length of a photon is half of the wavelength and the radius is proportional to square root of the wavelength. A photon can ionize a hydrogen atom at ground state only if its radius is less than the Bohr radius. *A photon and its copy have the phase difference of π and constitute a phase-entangled photon pair.* The photon may be considered a quark~anti-quark particle with all 3 colors, with its copy. A photon is in shape like a thin rod if its energy is lower than the rest energy of an electron and like a plate if its radius is smaller than the classical radius of an electron. An unaccelerated free electron can only emit one photon, when being annihilated by colliding with a positron. Both particles will be converted into gamma rays, one photon each, having an energy of approximately 0.51 MeV. Is the photon emitted away from the collision by the intrinsic parity of gluon pairs?

$$g^a = |r\ b\ g| \lambda^a \begin{vmatrix} \bar{r} \\ \bar{b} \\ \bar{g} \end{vmatrix} \tag{10.1}$$

Quark and gluon have colors. The quark color may be represented as an equilateral triangle, and the denotations $r, g, b, \bar{r}, \bar{g}, \bar{b}$, for colors and anticolors are chosen arbitrarily. In QCD, SU(3) Color Symmetry is provided by the strong interaction that is invariant under rotations in color space, $U = e^{-i\alpha_a \lambda^a}$, known as a non-Abelian symmetry, with the same strong interaction for all three colors.

Thus, for the quark states we have for the following colors:

$$r = \begin{vmatrix} 1 \\ 0 \\ 0 \end{vmatrix} \tag{10.2}$$

$$b = \begin{vmatrix} 0 \\ 1 \\ 0 \end{vmatrix} \tag{10.3}$$

$$g = \begin{vmatrix} 0 \\ 0 \\ 1 \end{vmatrix} \tag{10.4}$$

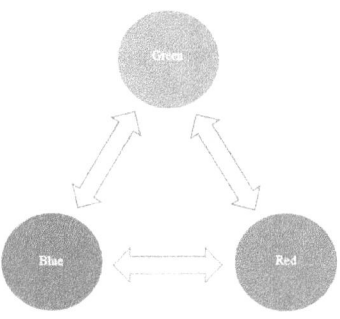

Figure 22. The Colors of the Force.

Experiments limit the gluon's rest mass to less than a few meV/c^2. The gluon has negative intrinsic parity. Parity is the flip in the sign of all three spatial dimensions. This condition supports the hypothesis that the electron may consist of 3 negative gluon~anti-gluon pairs.

Does the electron neutrino consist of pairs of gluons? For example, $r\bar{r} + b\bar{b} - 2g\bar{g}$. Is a neutrino an elementary particle with no substructure? or could it consist of gluons like the photon or electron?

Let us hypothesize that the electron neutrino is represented by $r\bar{r} + b\bar{b} - 2g\bar{g}$ that would be two pairs of gluons that might be represented by $r\bar{r} - g\bar{g}$ and $b\bar{b} - g\bar{g}$. These may be hypothesized to be virtual quarks. A quark~anti-quark pair forms mesons. Baryons are quark triplets. Mesons and baryons are hadrons. Bound states composed of quarks and gluons, are collectively known as partons.

How are these two pairs produced? According to recent experiments, a positive pair of red gluons and a positive pair of blue gluons of the same intensity have an affinity to combine with two negative pairs of green gluons. Could it be possible that the two negative pairs of green gluons $-2g\bar{g}$ have half the color intensity? It may be theorized that a photon $r\bar{r} + b\bar{b} + g\bar{g}$ may combine with a neutrino $r\bar{r} + b\bar{b} - 2g\bar{g}$ and produce a pair of green~anti-green gluons, $-g\bar{g}$, or an electron and a neutrino may combine to produce a negative triplet of green~anti-green gluons $-3g\bar{g}$ that may later recombine with a red or a blue positive pair of gluon~anti-gluon. In order for the neutrino to have a neutral color charge as theorized, the red and the blue color charges would have to be of the same intensity but the green color charge would have to be half of the intensity of either the red or the blue charges.

If these hypotheses are supported empirically, all the fermions or leptons, that are called elementary particles in the current standard model, have a gluon substructure. Then, even a boson, a force carrier, like a photon may have a gluon substructure. It is possible to suggest that if these particles can be generated from high energy collisions between photons, all particles have a common gluon origin as the exchange particle (or gauge boson) for the strong nuclear force and the spatiotemporal substance as suggested by A Dynamic Theory of Space-Time. Is it all a matter of waves and Bosons? (Nieves, 2020)

The intrinsic negative parity of gluons.

In quantum mechanics, a parity transformation is the flip in the sign of one spatial coordinate. In three dimensions, it can also refer to the simultaneous flip in the sign of all three spatial coordinates. The intrinsic parity is a phase factor that arises as an eigenvalue of the parity operation.

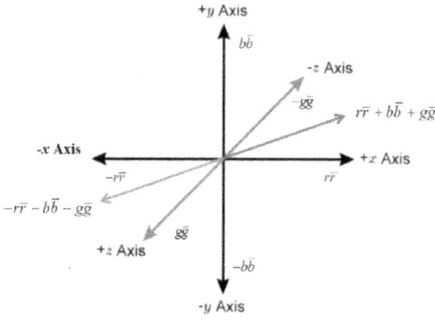

Figure 23. String Color-Spatiotemporal Cartesian Coordinate Axes.

In quantum mechanics, spacetime transformations act on quantum states. The parity transformation, $\hat{P}$, is a unitary operator, in general acting on a state Ψ as follows:

$$\hat{P}\psi(r) = e^{\frac{i\phi}{2}}\psi(-r) \tag{10.5}$$

The parity or space inversion operation converts a right handed coordinate system to left handed: $x - \rightarrow -x, y - \rightarrow -y, z - \rightarrow -z$. We consider the parity operation, i.e., we let the parity operator π to act on vectors of a Hilbert space and keep the coordinate system fixed, we have: $|\alpha\rangle - \rightarrow \pi/|\alpha\rangle$. Hence, negative parity may be interpreted as the changing of direction along a dimension from the retarded wave to the advanced wave at an arbitrary point. So, a gluon pair may change direction as it acquires negative parity or a spatiotemporal negative sign. If a photon changes direction as a result of negative parity it has the probability to become an electron in the opposite direction of its trajectory since each gluon pair has a negative charge of $-e/3$. Charge equals the product of a spatial length and a temporal length. As the gluon pairs travel in the advanced wave they acquire the temporal attribute of charge in addition to their spatial length.

$$\lambda_e = \frac{2(-r\bar{r} - b\bar{b} - g\bar{g})}{\sqrt[2]{6}} = \frac{1}{\sqrt[2]{3}} \begin{vmatrix} i^2 & i^2 & 0 & 0 & 0 & 0 \\ i^2 & i^2 & 0 & 0 & 0 & 0 \\ 0 & 0 & i^2 & i^2 & 0 & 0 \\ 0 & 0 & i^2 & i^2 & 0 & 0 \\ 0 & 0 & 0 & 0 & i^2 & i^2 \\ 0 & 0 & 0 & 0 & i^2 & i^2 \end{vmatrix} \tag{10.6}$$

$$\hat{P}\lambda_0(g) = e^{\frac{i\phi}{2}}\lambda_e(-g) \tag{10.7}$$

Parity has long been regarded as the third notion of symmetry, after attraction or repulsion in the polarity of charges, to be the invariance of physical laws and physical behavior when you look at them in a mirror. Since the early 1950s, it was shown for the weak nuclear force that there would be multiple violations of two of those symmetries. If you change the polarity of charges, and look in a mirror, there would be a breaking of symmetries. These breakings of symmetries have very serious implications for the microstructure of matter but they may also have implications for the macrostructure of the universe. (Yang, 1952)

Aside, every physical thing is free to be what it is and what it was made to do. A physical thing may manifest or direct itself. So every physical thing is contingent on what is already contingent. Some things we observe are the result of what we do not observe. The visible universe is part of the evidence and part of the result of what may not be visible.

Can a gluon be represented as a wave? Gluons exhibit particle duality. Quarks and gluons are confined within hadrons, they have a maximum wavelength in the order of the confinement scale. Gluons are said to be flavor imperceptive; that is, they don't distinguish between quark flavors.

After introducing color, the complete wavefunction of hadrons can now be written as:

$$\Psi_{Complete} = \Psi_{Space} \Psi_{Spin} \Psi_{Flavor} \Psi_{Color} \qquad (10.8)$$

Thus, the wave function of a baryon changes sign if two quarks are exchanged, as required by the Pauli principle. The Pauli principle states that the complete wavefunction must be antisymmetric under the interchange of any pair of identical fermions, and symmetric under the interchange of any pair of identical bosons.

Particles are classified into bosons and fermions, but in reality this is not so much about their spin, but more about the quantum identity of the particles.

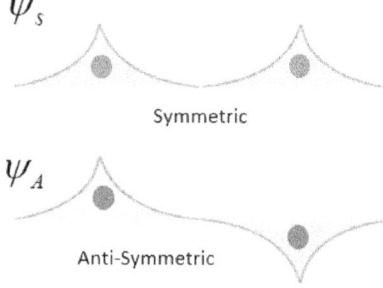

Figure 24. The Symmetry with Respect to Particle Exchange.

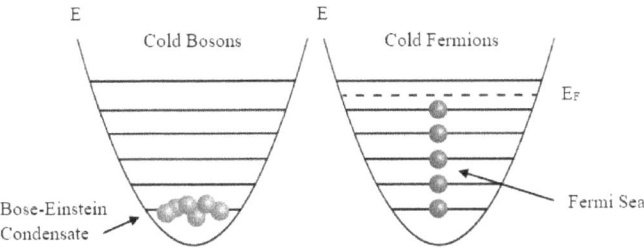

Figure 25. The Symmetry of Cold Bosons and Cold Fermions.

Likewise, the wave functions of the mesons are color singlets. Zero color charge means that the hadrons have the following color wave-functions:

$$\frac{1}{\sqrt{3}}(q\bar{q}) \to \frac{1}{\sqrt{3}}(r\bar{r} + b\bar{b} + g\bar{g}) \qquad (10.9)$$

The wave functions of the hadrons were assumed to be singlets of the color group. The baryon wave-functions are antisymmetric in the color indices, denoted by red (r), green (g) and blue (b):

$$\frac{1}{\sqrt{6}}(qqq) \to \frac{1}{\sqrt{6}}(rgb - rbg + brg - bgr + gbr - grb) \qquad (10.10)$$

The cross-section for electron–positron annihilation into hadrons at high energies depends on the squares of the electric charges of the quarks and on the number of colors. For three colors this leads to:

$$\frac{\sigma(e^+ + e^- \to hadrons)}{\sigma(e^+ + e^- \to \mu^+ + \mu^-)} \to 3\left[\left(\frac{2}{3}\right)^2 + \left(-\frac{1}{3}\right)^2 + \left(-\frac{1}{3}\right)^2\right] = 2 \qquad (10.11)$$

Without colors this ratio would be ⅔. The experimental data, however, were in agreement with a ratio of 2. All observed states (all mesons and baryons) have a total color charge that is zero. This is called color confinement.

Gauge bosons couple to conserved charges: *QED*: Photons couple to electric charges (*Q*). *QCD*: Gluons couple to color charges. Photons do not carry electric charge, but gluons do carry Quantum Chromodynamics color charges themselves!

Gluons do not exist as free particles since they have color charge. Gluons are massless, their frequency/wavelength/momentum/energy relations are the same as for photons. The gluon has no mass, and hence, gluons travel at the speed of light when created and annihilated in their exchange process within the nucleons. Hadrons contain only virtual gluons, which do not obey the ordinary relationships between energy and wavelength. High energy collisions create real gluons.

The photon structure function, in quantum field theory, describes the quark content of the photon. The function is defined by the process $e + \gamma \rightarrow e + hadrons$. This process has been derived from the experimental analyses of the photon structure function. The experiments utilize the so-called two-photon reactions at electron–positron colliders $e^-e^+ \rightarrow e^-e^+ + h$, where h includes all hadrons of the final state. Hence, the photon consists of gluons.

If an electron may consist of gluon pairs, why would nature not use a combination of gluon pairs to create an electron or any lepton heavier than an electron? Do fermions and leptons consist of gluon pairs? An electron can decay into a photon, and a photon consists of gluons. An electron is fundamental because they are all non-unique. Electrons are generally thought to be elementary particles because they had no known components or substructure. Nonetheless, the faster an electron is measured experimentally in a large particle accelerator, the amount that is measured is greater, through the scaling of charge, of the long distance scale property, a long wavelength, of space-time, that increases to the point of the grand unification scale and attains the same magnitude as the strong nuclear interaction. Hence, gluons could be that substructure. Could an electron represent a fundamental generation of gluon pairs with a negative charge due to parity?

$$gluons \rightarrow -\gamma \rightarrow e^- \qquad (10.12)$$

If an unaccelerated electron gets annihilated by a positron, each lepton is converted into a photon of a gamma ray. The momentum of the impact of the matter~anti-matter collision reverses the physical orientation of each gluon pair in the electron that causes symmetry breaking or parity.

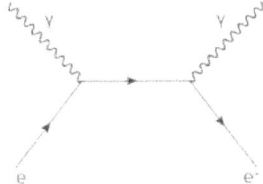

Figure 26. The Annihilation of an Electron with a Positron.

$$-r\bar{r} - b\bar{b} - g\bar{g} \rightarrow \bar{r}r + \bar{b}b + \bar{g}g \rightarrow r\bar{r} + b\bar{b} + g\bar{g} \quad (10.13)$$

It is possible to theorize that as the symmetry of each gluon pair in an electron breaks during parity, the gluon field of each pair reverses itself, traveling in the opposite direction as a photon or carrier of the electromagnetic force. Thus, the photon travels in the direction of the electromagnetic field, but the electron travels in the opposite direction to the electromagnetic field toward the positive charges of the field potential.

Gluon Standard Model	$\bar{r}$	$\bar{b}$	$\bar{g}$
r	$\pm r\bar{r}$	$\pm r\bar{b}$	$\pm r\bar{g}$
b	$\pm b\bar{r}$	$\pm b\bar{b}$	$\pm b\bar{g}$
g	$\pm g\bar{r}$	$\pm g\bar{b}$	$\pm g\bar{g}$

Figure 27. The Gluon Standard Model.

The most common gluon color is red, followed by blue, and then green. The symbol "±" represents the intrinsic Parity of a gluon pair. All the particles of the current standard model of particle physics (quarks, leptons, and bosons that mediate forces) may be represented by the Gluon Standard Model. The substructure of all particles of mass is very colorful. It is possible to theorize that gluons are the fundamental colors of all particles or systems of mass. The strongest quarks are between two or three gluon pairs of the same color. A doublet is the most common with red being the most abundant color in the model. A triplet may be a photon, a lepton, or a boson. A doublet may be any type of quark. A triplet or doublet may interact with other triplets or doublets, creating or canceling other triplets or doublets. The most stable and common doublets consist of a (<u>color</u>~anti-color) ± (color~anti-<u>color</u>) between two different colors.

The main diagonal of the gluon color matrix consists of those elements that lie on the diagonal that runs from the top left to the bottom right. The minor diagonal intersects the main.

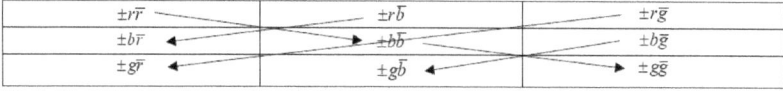

$\pm r\bar{r}$	$\pm r\bar{b}$	$\pm r\bar{g}$
$\pm b\bar{r}$	$\pm b\bar{b}$	$\pm b\bar{g}$
$\pm g\bar{r}$	$\pm g\bar{b}$	$\pm g\bar{g}$

Figure 28. The Stable Color States.

Gluon Pairs
$r\bar{r} \pm b\bar{b}$
$b\bar{b} \pm g\bar{g}$
$r\bar{r} \pm g\bar{g}$
$b\bar{b} \pm r\bar{g}$
$b\bar{b} \pm g\bar{r}$
$r\bar{g} \pm g\bar{r}$

Figure 29. A List of Potential Quark Doublets.

From previous research, it is possible to theorize that the above potential quark doublets from the gluon color matrix may recombine to generate photons, leptons, and force-mediating bosons. There may be gluon pairs remainders that will further combine with other free gluon pairs, will be interchanged to change a quark flavor, or become part of force-mediating bosons. From the above list of potential quark doublets, the color gluon pair, $r\bar{r} - b\bar{b}$, and $r\bar{g} \pm g\bar{r}$, remain as stable color states. The triplet, $r\bar{r} + b\bar{b} - 2g\bar{g}$, a stable color state, may be derived from the list of potential quark doublets.

The Gluon Parity Effect

From previous research, the structure of the photonic field is usually represented by two components, the electric field, and the magnetic field, at quadrature. A photon may have a linear or circular polarization. A rotating electric field or a rotating magnetic field may originate from a rotating dipole of charges within a conductor. However, the photonic field may be represented by a third field which may be considered a triadic electromagnetic field. The photon has been described as a quantized bundle of electromagnetic energy that acts like a particle, with no rest mass, traveling at the speed of light, which has momentum. The quantized energy of the photon is given by the Planck constant times the frequency of the photon. (Nieves, 2021)

The magnetic field is the projection of the photonic field on the spatial plane of space-time, a spatiotemporal pressure field. The photon is a corpuscle of pure energy density (a light quantum) that creates spatiotemporal pressure around itself, curving the spatiotemporal medium in the process. The electric field is the projection of the force exerted by the energy density on the temporal plane of space-time along the direction of propagation, a directional force. The energy density of the light quantum travels on the temporal wave capable of exerting a luxonic force on a charged particle in its trajectory. (Nieves, 2020)

The electric field exerts the force, N/Q, of the relativistic mass of the photon and its pressure at the speed of light, through the spatiotemporal medium onto the charges of particles and objects of mass and energy. The electric tube of force applies a directional force on an area of the atomic lattice, and through the plenum, of an object of mass in the direction of propagation.

The magnetic field exerts lateral pressure, N/m^2, on, but not limited to, the spatiotemporal medium, and other parallel tubes of force. As the photonic wave follows its path from the North pole of a permanent magnet to the South pole, some tubes of force extend through the regions of highest spatiotemporal pressure near the mass of the magnet, and the subsequent tubes of force extend at the closest possible parallel distance to the contiguous tubes, as all the tubes of force build the magnetic field through curved space-time. It is possible to hypothesize that as a photon travels in a magnetic field, it perturbs the spatiotemporal medium to create a tube of force with higher pressure along its trajectory, which becomes a spinning magnetic tube and a structured magnetic conduit for the next photon. If two opposite poles of two permanent magnets attract, the tubes of force create a direct electromagnetic circuit, so the outside poles become the opposite poles, or the North and South poles, of the overall electromagnetic field of the composite magnet.

Consequently, as the electrons flow in an electrical circuit opposite to the electric field toward the anode, the electrons (negative charges) would be conveyed through the electric tubes of force within the conductor by a spatiotemporal pressure differential that extends from anode to cathode.

What is an electromagnetic field made of?

The electromagnetic field is closely associated with the photon. There are two types of electromagnetic fields: waves and quasi-static fields. The quasi-static electromagnetic fields consist of virtual photons. The electromagnetic waves consist of real photons.

The real photons characterize the behavior of the electromagnetic waves. The real photons have two independent polarization states, or a non-unique choice of basis. The basis is a full mathematical method. The two independent states of real photons are transverse. The virtual photons characterize the behavior of quasi-static electromagnetic fields. The two independent states of virtual photons are longitudinal.

The electromagnetic interaction is mediated by the constant exchange of photons from one charged particle or object to another. Electromagnetic interactions may involve real photons with definite frequencies, momenta, and energies, or the exchange of virtual photons may be involved with the electrostatic and magnetic fields. Thus, it is possible to suggest that there may be a dense cloud of virtual photons very near an electron that may be emitted and re-absorbed by the electron. Some of these virtual photons may split into electron-positron pairs, that recombine into virtual photons that are re-absorbed by the same electron. These virtual photon circuits shield the charge of the electron, so that from far away, the electron seems to have less charge than nearby.

Let us consider an ideal capacitive circuit where the capacitor is fed intermittently from its source in a perfect vacuum without any external source of electromagnetic radiation. During the disconnection period of the source to the capacitor, the capacitor is connected to a load through a switch, where its charges (electrons) flow through the conductors from the cathode to the anode of the capacitor. At the instant that the capacitor is fully charged, the very thin cathode plate has a uniform distribution of electrons. There is a full electric field from the anode to the cathode that attracts the negative charged electrons. At the junction of the cathode plate with the spatiotemporal medium between the plates, as a surface electron goes through the quantum process of gluon parity, where the gluon pairs flip due to the electric field potential between the plates to become a photon, the photon is propelled toward the anode of the capacitor to manifest an electric field. As the photon arrives near the junction of the spatiotemporal medium with the anode, the photon may go through the reverse quantum process of gluon parity to become an electron on the anode plate as the anode would eventually fill up with numerous electrons that flow through the circuit conductors. Since the anode lacks electrons, the probability of the photoelectric effect occurring may be negligible.

The electromagnetic field is also manifested through the conductors and about the conductors as the electrons flow through the conductive medium, with the electric field flowing from anode to cathode, as the magnetic field flows about the conductor by Fleming's righthand rule.

It is possible to suggest that when a photon affects an electron in an atom, whereby the electron may absorb some of the energy in the incident photon to change its orbit or be ejected from its atom, the photoelectric effect may be causing the onset of the quantum process of gluon parity at the atomic level. An electron may also be manifested in an atom through the doubling of gluon pairs increasing the overall color charge of the resulting electron.

§ 11. The Color String Theory.

The representation of a gluon state as a color corpuscle and a wave may provide an interesting frequency analysis of the substructure of particles and systems of mass. The color states, the colorness intensity, and the amount of mass gain by a corpuscle are proportional. (Nieves, 2020)

The energy "E" of colorness for color or anti-color charge(s), may be expressed as

$$E = \pm k\vec{g}d = \pm \bar{\lambda}\omega_p \vec{g}d \tag{11.1}$$

where the lambda-bar, $\bar{\lambda}$, is the color temperature quantum of action, or reduced color temperature constant, for a pixel, or an anti-pixel, with colorness, $\pm k$, ω_p is the associated angular frequency of vibration in Planck units, $\vec{g}$ is the gluonic field between colors and anti-colors, in Newtons/±colorness, and d is the distance between a color and an anti-color.

What would be the gluonic acceleration "$\ddot{g}$" about a color charge in a homogenous and isotropic medium?

$$\ddot{g} \equiv \frac{\partial \vec{g}}{\partial m'} = \hat{G}(\pm k_p)r^2 \tag{11.2}$$

$$\hat{G} = \frac{c^2}{\pm k \cdot a} \tag{11.3}$$

where $\vec{g}$ is the gluonic field, m' is the relativistic virtual mass of the force of a gluonic field, $\hat{G}$ is the Gell-Mann quantum colorness constant, $\pm k_p$ is the colorness of pixels or antipixels, "r" is the distance from the center of colorness, "c" is the speed of light, and "a" is the spatial volume.

A theory of color strings may effectively describe the interactions of all the electromagnetic forces of the current standard model. The gluonic or gluon fields function like strings holding together the quarks with some tension. The modes of oscillation of these color strings could potentially describe all the particles within the quantum framework of the current standard model, and is applicable to algebraic geometry, black hole physics, cosmology, and condensed matter physics. Under the hypothesis that space-time is fundamental and emergent, color string theory starts at its successful origin of development where it is a theory for gluons and their substructure of colorness, without having space-time quantized for the sake of mathematical efficacy, but for, and not limited to, the energy and fields of gluons, as well as other characteristics, that remain as parts of a theory of quantum gravity. Even if space-time has been theorized in previous research as quintessential and the source of all there is, color string theory may play a crucial role as a foundational theory from which other successful theories of existing mass and systems of mass and their related physical fields, may be derived and empirically verified. The central premise of the theory of color strings is that color strings are embedded in space-time. Color strings consist of the spatiotemporal substance and exist in the fundamental and emergent spatiotemporal background. The color strings are spring-like and may stretch out as strings of the strong nuclear force.

All particles and carriers of a force may be expressed in terms of oscillations in color strings. The color string waves may interfere constructively or destructively. A color theory that encompasses supersymmetry, a theoretical symmetry between bosons and fermions, can minimize the number of dimensions involved, and explain the vast difference in strengths between fundamental forces. Hence, the theory of color strings will be referred as color string theory from this point forward within this document.

Let us propose a color string theory as a theoretical framework in which the point-like pixels of six-dimensional chromodynamics are replaced by one-dimensional objects called color strings. The color string theory would describe how these color strings propagate through space and time to interact with each other. Hence, color string theory proposes that the fundamental constituents of the universe are one-dimensional "color strings" rather than point-like pixels. The color string theory proposes three dimensions of space, three dimensions of time, three dimensions of electromagnetic charge, a zero-dimension for a spatiotemporal point, a point-like pixel or a source of space-time-color, and it contains ways of relating expanded spatiotemporal dimensions to contracted spatiotemporal dimensions. Color string theory does not promise to provide a way to unite General Relativity and Quantum Mechanics because it has been found in previous research that those two theories are part of a Synthesis of Quantum Theory for six-dimensional space-time. (Nieves, 2021)

The proposed dimensions of color string theory previously mentioned may also be conceptualized as three spatial coordinate dimensions (x, y, and z), three temporal coordinate dimensions $(ct_x, ct_y,$ and $ct_z)$ that are spatially comparable, and three electromagnetic dimensions of charge that may be expressed as the product of three additional spatial dimensions and three temporal dimensions. The temporal dimensions may be folded into a resultant temporal dimension for a (9 + 1) formalism of nine space-like dimensions and one resultant time-like dimension for mathematical efficacy. If a zero dimension is also included the above color string theory would represent a M-color string theory with a (10 + 1) formalism for a single framework of color string theory with minimal dimensions.

If in principle, the gluon standard model lives somewhere in the color string landscape, then the geometry of the dimensions involved must be verifiable and testable predictions could be made beyond the gluon standard model. (Nieves, 2020)

Color string theory goes beyond the current description of the universe by replacing all matter and each force carrier with just one element: a quantum-size vibrating color string that twists and turns in complicated ways that, from a macroscale perspective, would look like a point-like pixel. Would color string theory be a literal theory of colorness, a single unifying framework that explains all the variety and hues of colors that are present in the theory of chromodynamics of the micro universe in quarks, hadrons, and fundamental particles, to why leptons have the mass that they do?

Would color string theory represent a description of all forces and matter in one mathematical expression? Color strings may collide and rebound cleanly without implying physically impossible infinities. A one-dimensional color string that really tames the infinities that may come up in the calculations.

An extended color string of a length of a particular color-anticolor triplet vibrates in a particular frequency that has the properties of a photon, or a lepton, and another extended color string of a particular color-anticolor doublet that folds and vibrates with a different frequency that has the properties of a quark, etc.

Color string theory is highly constrained. It is only dependent on a one-dimensional parameter, the color string length ℓ_s.

Even though, it may be expected that the spatial, temporal, and energy scales, involved in color string theory may take values close to unity in Planck units. Hence, the color string length may be expressed in approximately the same order as the Planck length.

Moreover, the spatiotemporal dimensions are uniquely verifiable to be ten for the internal mathematical consistency of the theory. Also, one vibrational mode of closed color strings corresponds to the graviton, so quantum gravity is an expected consequence of the color string theory.

In color string theory, a gluon brane is a physical object that generalizes the notion of a point-like pixel to higher dimensions. Gluon D-color branes may be dynamical objects which can propagate through space-time according to the rules of six-dimensional chromodynamics and quantum mechanics.

The properties of an electromagnetic charge may provide additional dimensions for gluon D-color branes in the macroscale. According to six-dimensional Chromodynamic principles, these properties are derived mostly from color charges and gluonic fields.

A Gluon Dirichlet Color Brane is an object extended in one or more spatiotemporal dimensions, which arise in color string theory and in a Synthesis of Quantum Gravity. (Nieves, 2021)

A D0-color brane is a zero-dimensional point-like pixel; a D1-color brane is a one-dimensional color string; a D2-color brane is a two-dimensional color membrane; and a p-color brane is a p-dimensional color object.

It is possible to suggest that color string theory may be considered a frequency string theory where the color string or D0-color brane may be the fundamental frequency or source of space-time-colorness, and other D#-color branes are harmonics of the fundamental. So, each color gluon would have its own unique frequency range.

Gluon Frequency	$\omega_{\bar{r}}$	$\omega_{\bar{b}}$	$\omega_{\bar{g}}$
ω_r	$\pm\omega_{r\bar{r}}$	$\pm\omega_{r\bar{b}}$	$\pm\omega_{r\bar{g}}$
ω_b	$\pm\omega_{b\bar{r}}$	$\pm\omega_{b\bar{b}}$	$\pm\omega_{b\bar{g}}$
ω_g	$\pm\omega_{g\bar{r}}$	$\pm\omega_{g\bar{b}}$	$\pm\omega_{g\bar{g}}$

Figure 30. The Gluon Frequency Standard Model.

There may be several types of gluon D-color branes, including the fundamental color strings whose quantization defines color string theory; black color branes, which may be solutions to the EFEs that may resemble black holes but are extended in some dimensions rather than being spherical; and there may be D-color branes, which have the distinctive property that fundamental color strings can end on them with the end points of color strings stuck to the D-color brane or gluon-ball.

A gluon-ball, or glueball, is a hypothetical composite D-color brane. It consists solely of gluon color strings. Such a state is possible because gluon color strings carry color charge and experience the strong interaction between themselves. Color string objects in color string theory may need to vibrate in more than just the three spatial dimensions of space.

In a color string theory with gluon D-color branes, matter might be glued on a D-color brane that is embedded within the six-dimensional spatiotemporal dimensions. This improves the possibilities for understanding the laws of physics in terms of the six-dimensional spatiotemporal geometry. A consequence of this may be that the spatiotemporal dimensions may be expanded or contracted.

Gluon D-color branes may also appear in some of the cosmological inflation models of the early universe. As inflation requires a source of vacuum energy, that may be supplied by the rest mass of gluon D-color branes, the transition from inflation to ordinary expansion may be understood from the decay of gluon D-color branes into ordinary matter and radiation.

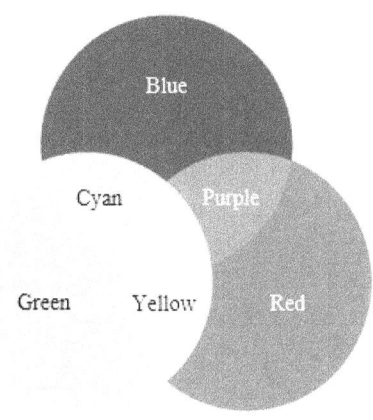

Figure 31. A Palette of Gluons.

§ 12. Are color strings fundamental perturbations with topological properties in the spatiotemporal fabric of physical reality?

A color string may be conceptualized as an oscillation of a resultant spatiotemporal wavefunction between two arbitrary points at the Planck scale. The distance between the points would be the length of the color string. The oscillation of the quantum wavefunction through its expansion or contraction would be the frequency of vibration of the color string or color brane. An arbitrary spatiotemporal point in situ may also oscillate through its expansion or contraction. It is theorized that the oscillations, torsions, and spatiotemporal pressures of color strings determine the properties of the particles that form gluonic and quantum substructures, and as a result atomic and classical structures.

At an arbitrary point along one spatiotemporal dimension, there are two wave functions, the retarded wave and the advanced wave. A color string "ℓ_s" has both the retarded wave and the advanced waves interfering within its length which gives the color string its fundamental frequency and its unique colorness. The temporal expansion of a one-dimensional color string "ℓ_s" would result in a two-dimensional color brane "D_2" that has its two-dimensional wavefunction oscillating to give the color brane its harmonic frequency and its unique colorness.

It is possible to suggest that any *p*-dimensional color object may oscillate into higher dimensions than its own. For example, a one-dimensional color string may oscillate into a second dimension that is perpendicular to its length. The sum of all harmonic frequencies of all related color branes add up to the fundamental frequency of the fundamental color string. A color string loop may form around an arbitrary spatiotemporal point due to the local curvature where the two ends of the color string may join together to manifest a color graviton. The retarded and advanced waves would oscillate and travel in opposite directions of the loop which would manifest an instant of gravity. The color graviton may propagate through other dimensions in expanding or contracting space-time. The color string loop or color graviton would still have its colorness and frequency that would follow the law of attraction between color and anti-color.

The retarded wave may be represented by a "ket", $|\Psi^+\rangle$, with complex spatial coefficients of the spatial basis vectors, and the advanced wave may be represented by a "bra", $\langle\Psi^-|$, with complex conjugate temporal coefficients of the temporal basis vectors. At an arbitrary spatiotemporal point, we may represent the advanced and retarded waves in bra-ket notation, $\langle\Psi^-|\Psi^+\rangle$, in six-dimensional space-time. A Spatiotemporal Density Operator $"|\psi^+\rangle\langle\psi^-|"$ is a density matrix for the multiplication of a retarded wave with its advanced wave at an arbitrary spatiotemporal point. A density matrix describes the statistical spatiotemporal state of a pure or a mixed quantum mechanical system. (Nieves, 2021)

We may express the color string in terms of its wavefunction as

$$\ell_s \equiv |\psi^+\rangle\langle\psi^-| \tag{12.1}$$

If we consider a rigidly rotating open color string, the length of the color string "ℓ_s" may be denoted as:

$$\ell_s = \frac{J}{p} \tag{12.2}$$

The color string length is given by the ratio of the angular momentum "J" in $\left(Kg \cdot m^2/s\right)$ and the linear momentum "p" in $\left(Kg \cdot m/s\right)$ during a fixed temporal instant "τ". For a given color string only certain frequencies corresponding to certain energies are possible. These resonant frequencies depend on the length of the color string.

Thus, color string length defines mass, complex vibrational modes, which in turn define particle properties like electromagnetic charge, gravitation, and spin. A single parameter "color string length" may define other crucial parameters in color string theory as a fundamental theory.

The color graviton may be defined as

$$g_s \equiv \pm m'_s \omega_s^2 \ell_s^2 \qquad (12.3)$$

where "ω_s" is the angular frequency of the color graviton, and "m'_s" is the relativistic mass. The angular frequency of a color graviton may be "$\pm \omega_s$" depending on whether the graviton is attractive of repulsive.

The color string tension "T_0" is given by

$$T_0 = \frac{\hbar c}{2\pi \ell_s^2} \qquad (12.4)$$

where "T_0" is given in Newtons, and "$\hbar$" is the reduced Planck constant. The tension of a color string defines its wave velocity also relates its frequency to its wavelength. Particle mass may be gained from the length of the color string and its tension "E/ℓ_s".

Color strings can expand, contract, vibrate, and hold energy. These properties provide a mechanism for color strings, or color branes, to interact and to decay into other elementary particles. The intrinsic abilities of one-dimensional color strings to join together or split apart gives color string theory the capacity to quantize gravity. The color graviton is a loop, not a spherical object, and its interactions are spread around the color string, avoiding the emergence of mathematical infinities. Color gravitons may emerge from the source of space-time-color as a color loop with color strings or color branes.

The slope parameter (α') is a fundamental constant that describes hadrons (quarks, gluons, and the strong nuclear force) that may be expressed as

$$\alpha' = \frac{J}{\hbar E^2} \tag{12.5}$$

In terms of the energy of the colorness and the gluonic field between color and anti-color charges, we have

$$\alpha' = \frac{J}{\hbar \left(\vec{kgd}\right)^2} \tag{12.6}$$

A rigidly rotating open string has a unique quadratic relation between its angular momentum "J" and its energy "E". Hence, for the color string wave behavior, only specific discrete energy modes are allowed. Those discrete vibrational modes can match the properties of known particles.

A relativistic color string traces out a worldsheet in space-time.

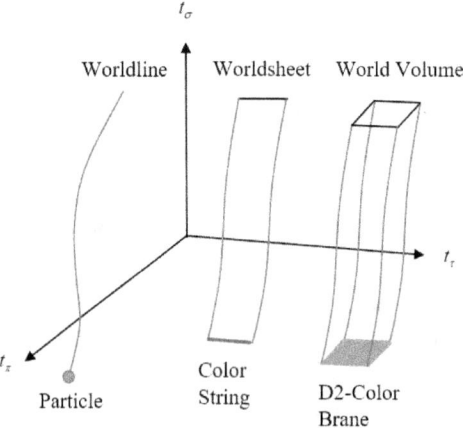

Figure 32. The world traces of relativistic color objects.

The color string worldsheet is a two-dimensional manifold with temporal parameters "τ", "π" and "σ" for a temporal representation of the color string at a fixed temporal instant "τ". For example, one end of the color string may be at "σ" and the other end is at "τ".

The string worldsheet in (d+3) spatiotemporal dimensions is thus described by the set of points:

$$\chi^0(\tau,\sigma,\pi), \chi^1(\tau,\sigma,\pi), ..., \chi^d(\tau,\sigma,\pi) \qquad (12.7)$$

where χ^d are the color string spatial coordinates and $\tau, \sigma,$ and π are the temporal coordinates.

The above (d+3) formalism describes two unique string types that follow the following specific conditions:

a. Open strings that may be periodic in the variable "σ".

b. Closed strings may be periodic in the variable "σ", so for any fixed time "τ" the closed color strings obey $\vec{\chi}(\tau,0,0) = \vec{\chi}(\tau,\sigma,0)$.

The center-of-mass motion of the color string and its oscillation about the center of mass can describe the motion of the relativistic color string as the color string obeys its wave equation of motion.

The following classical form provides the solution for the coordinates of the color string when the open relativistic color string is quantized.

$$\chi^\gamma(\tau,\sigma) = x_0^\gamma + 2\alpha' p^\gamma \tau + i\sqrt{2\alpha'} \sum_{n=1}^{\infty} \left(\alpha_n^\gamma e^{-\text{int}} - \alpha_{-n}^\gamma e^{\text{int}}\right) \frac{\cos n\sigma}{\sqrt{n}} \qquad (12.8)$$

On the right side of the classical form, the terms $\left(x_0^\gamma + 2\alpha' p^\gamma \tau\right)$ describe the position x_0^γ of the center-of-mass of the color string and the velocity of the center-of-mass of the color string, via the momentum p^γ in the $\gamma - th$ direction, and the time coordinate "τ".

The oscillation of the color string about its center of mass is described by the summation over oscillatory terms. The coefficients α_n^γ and α_{-n}^γ essentially correspond to Fourier coefficients in the classical description of the color string, as the summation corresponds to a Fourier series giving the contribution of each mode to the color string oscillation. When the relativistic string is quantized, the terms α_n^γ and α_{-n}^γ represent annihilation and creation operators that destroy and create excitations of each mode, analogous to a quantum harmonic oscillator. (Zwiebach, 2009)

The closed color string has a solution that is similar to the open color string for the coordinates of the color string, even though the oscillatory terms reflect the periodicity of the closed color string in a slightly different way.

$$\chi^\gamma(\tau,\sigma) = x_0^\gamma + \alpha' p^\gamma \tau + i\left(\sqrt{\frac{\alpha'}{2}}\right)\sum_{n\neq 0}\frac{e^{-in\tau}}{n}\left(\alpha_n^\gamma e^{in\sigma} + \tilde{\alpha}_n^\gamma e^{-in\sigma}\right) \quad (12.9)$$

The operator α_n^γ creates and destroys oscillation moving to the left while the operator $\tilde{\alpha}_n^\gamma$ creates and destroys oscillations moving to the right, for the closed color string, with $n > 0$ annihilating excitations and $n < 0$ creating excitations. The closed color string periodicity requires that the oscillations to the left are equal to the oscillations to the right. The additional term comes from that fact that the zero modes $(n \neq 0)$ may not be equal, $\tilde{\alpha}_0 - \alpha_0 = \sqrt{2\alpha'}\omega$.

The following equation for the square of the mass from the Hamiltonian of the closed color string in the quantum gravity theory includes the covariant number operators N and $\tilde{N}$ of the oscillations.

$$M^2 = \frac{2}{\alpha'}(N + \tilde{N} - 2) \quad (12.10)$$

The operator numbers for the left or right oscillations are integers. The scope of the mass is discrete which supports color string theory as a crucial part of a quantum gravity theory. If the operator numbers are equal to 1, the square of the mass vanishes.

An arbitrary color string state Ψ_s is a linear combination of two single creation operators for left and right movements in different color string coordinates $(\chi^\gamma, \chi^\lambda)$ acting on the spatiotemporal volume.

$$\Psi \sim \sum R_{\gamma\lambda} \alpha_1^\gamma \alpha_{-1}^{-\lambda} |0\rangle \qquad (12.11)$$

where $R_{\gamma\lambda}$ is an arbitrary matrix of the appropriate size.

The arbitrary matrix $R_{\gamma\lambda}$ may be decomposed into its traceless-symmetric part $S_{\gamma\lambda}$, an antisymmetric part $A_{\gamma\lambda}$, and a trace part $T\delta_{\gamma\lambda}$.

$$R_{\gamma\lambda} = S_{\gamma\lambda} + A_{\gamma\lambda} + T\delta_{\gamma\lambda} \qquad (12.12)$$

The scope of the closed color string in color string theory contains massless states whose degree of freedom are carried by a traceless symmetric matrix $S_{\gamma\lambda}$. So, the closed color string scope provides a quantum description of classical graviton states. The classical description of one-graviton state in General Relativity is exactly the same. The other degrees of freedom of the arbitrary matrix $R_{\gamma\lambda}$ are also important in the context of color string theory. The antisymmetric part $A_{\gamma\lambda}$ corresponds to the Kalb-Ramond field which gives color strings a type of electric charge. The trace part $T\delta_{\gamma\lambda}$ corresponds to a massless scalar field called the color dilaton.

A color dilaton is a field on lower spatiotemporal dimensions which is a component of the quantum gravity field on the higher spatiotemporal dimensions, in that it is part of the color metric of the color fiber-spaces on which a spatiotemporal compactification takes place. The expectation value of the color dilaton field sets the value of the color string coupling constant and therefore governs the strength of color string interactions. In special relativity space-time, the relativistic point particles develop in a manner that the length of the particle worldline is extremized. Comparably, the relativistic color string develops in a manner that the area of the color string worldsheet is extremized. The following action for the color string coordinates may be expressed using differential geometry as the

Nambu-Goto action, a functional that encodes the area of the color string worldsheet.

$$S = -\frac{T_0}{c} \int d\tau d\sigma \sqrt{\left(\frac{\partial \vec{X}}{\partial \tau} \cdot \frac{\partial \vec{X}}{\partial \sigma}\right)^2 - \left(\frac{\partial \vec{X}}{\partial \tau}\right)^2 \left(\frac{\partial \vec{X}}{\partial \sigma}\right)^2} \qquad (12.13)$$

The above action is, up to a constant, equal to the area of the color string worldsheet. The constant T_0 has units of spring tension. The Polyakov action in color string theory may be considered physically equivalent. The equations of motion for the color string coordinates from the Nambu-Goto action are the wave equation in each coordinate for an expressly simple form of the parameters (τ, σ). The color string motion is precisely oscillatory since the color string coordinates obey the wave equation.

$$\frac{\partial^2 \vec{X}}{\partial \sigma^2} - \frac{1}{c^2} \frac{\partial^2 \vec{X}}{\partial t^2} = 0 \qquad (12.14)$$

The boundary conditions are provided below for the color string endpoints, and for the periodicity of the closed color strings. There are two types of boundary conditions for open color strings, and one periodic boundary condition for a color string loop.

For Dirichlet boundary conditions:

$$\left.\frac{\partial \vec{X}}{\partial \tau}\right|_{\sigma=0} = \left.\frac{\partial \vec{X}}{\partial \tau}\right|_{\sigma=\sigma^*} = 0 \qquad (12.15)$$

For Neumann boundary conditions:

$$\left.\frac{\partial \vec{X}}{\partial \sigma}\right|_{\sigma=0} = \left.\frac{\partial \vec{X}}{\partial \sigma}\right|_{\sigma=\sigma^*} = 0 \qquad (12.16)$$

For Sweetchild periodic boundary conditions:

$$\frac{\partial \vec{X}(\tau + \pi, \sigma)}{\partial \pi} \equiv \frac{\partial \vec{X}(\tau, \sigma)}{\partial \pi} \qquad (12.17)$$

Under Dirichlet boundary conditions, the ends of the color string are fixed to some physical object. Under Neumann boundary conditions, the ends of the open color strings are always flat and do not feel a force. The Neumann boundary conditions are also known as free-endpoint boundary conditions.

The Sweetchild periodic boundary conditions imply that the ends of the color string are always joined forming a color string loop. So, we may have a color string loop with the following condition, $\vec{\chi}(\tau=\pi,\sigma)=\vec{\chi}(\tau=0,\sigma)$. As the two end points are identified, the string becomes a topological loop, and its worldsheet is cylindrical. All the boundary terms $B\big|_{\tau}^{\pi}$ vanish simply because the values at $(\tau=\pi\ number)$ cancel against those at $(\tau=0)$. If $\vec{\chi}(\tau,\sigma)$ is decomposed into its Fourier modes, the most natural modes would be like $e^{2in\tau}$, where "n" is an integer, and there would be left-moving complex waves as well as right-moving complex waves on the world sheet. A color string loop can propagate anywhere in the bulk and may be considered the representation for a graviton. Currently, it is thought that most of the mass of elementary particles comes from gluons and their fields.

The physical objects for these conditions are D-color branes which are excitations related to specific states of open or closed color strings. A p-dimensional D-color brane with endpoints that are restrained is a Dp-color brane.

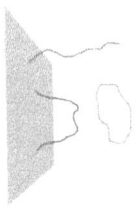

Figure 33. Types of Attachments for a D-color Brane.

Let us express six-dimensional Color String Field Equations in terms of the natural law of pressure equal to energy density using the spatiotemporal radial curvature operator on the surface area of the D2-color brane, $\Gamma_{SP}^2 D_2$, for a complex wave function.

$$\Gamma_{SP}^2 D_2 g_{\mu\nu} = \frac{4\pi G}{c^4} T_{\mu\nu} \qquad (12.18)$$

where $T_{\mu\nu}$ is the color string energy density tensor for the surface area, $g_{\mu\nu}$ is the metric of the spatiotemporal medium of a D2-color brane, and ρ is the energy density, $\rho \equiv \pm k\ddot{g}d / dA$, of an arbitrary D2-color brane.

From previous research, the six-dimensional Color String Field Equations represent the equivalence between pressure and energy density for a triangular surface area of curvature of a D2-color brane with respect to a spherical surface area with a radius "r". (Nieves, 2021)

Aside, it is interesting to note that the color quantum charge of a D2-color brane "q_b" may be expressed as the product of the spatial area of a color brane "ℓ_b^2" and the square of a temporal period "t_b^2". So, let us visualize a D2-color brane folded into the shape of a hollow cylinder. A two-dimensional color brane wrapped around other dimensions will look like a cylinder.

Thus, using a Planck rectangular coordinate system, "q_b" may be written as

$$q_b \equiv 2\pi \sqrt{\left(\ell_b^2 \cdot t_b^2\right)} \qquad (12.19)$$

$$q_b = \sum_{p=1}^{n} \left(a_p \cdot q_p\right) \qquad (12.20)$$

Where "n" is the maximum number of color quantum charges on an arbitrary D2-color brane, and "a_p" is the dimensional coefficient of the geometry at or around a point, typically 2π for one spatial and one temporal dimension of a length or for two spatial and two temporal dimensions of the external surface of a hollow cylinder, and 4π for two spatial and two temporal dimensions of a spherical surface. For an open D2-color brane, "a_p" would be equal to "1". Hence, each spatiotemporal plane of a rectangular coordinate system may be defined with three additional dimensional axes of charge, that are spatiotemporal in nature, as previously suggested. A hypothetical color dilaton particle has been defined as a particle of a scalar field "ϕ" that appears in multidimensional theories when the volume of the compactified dimensions expands or contracts.

A color string loop, that expanded as a color dilaton, may be expressed as a color string loop quantum charge, given by

$$q_o = 2\pi r_s t_s \qquad (12.21)$$

Currently, the Higgs boson is the fundamental particle associated with the Higgs field, a field that gives mass to other fundamental particles such as leptons and quarks. The mass of a particle determines its inertia when it encounters a force.

The Higgs particle field creates a particular density of energy permeating the universe. In the current Standard Model, the Higgs particle is a massive scalar boson with zero spin, no electric charge, and no color charge. What if the Higgs consisted of color string charges since all elementary particles and bosons are theorized to derive from color string charges?

It is now a good time to ask the following rhetorical questions. Does the Higgs field consist of color string loops, or color gravitons? A color string loop, or color graviton, can propagate anywhere in the spatiotemporal dimensions of physical reality. If the color string loops were attracted to other open color strings or D-color branes, the mass near the color Planck-size objects, or near the elementary particles or bosons, would increase accordingly. It is theorized that if two color string loops interact, there would be a process by which two color string loops would join into an intermediate color string loop which splits apart into two color strings loops again. Would color string loops be able to form open or closed chains of color gravitons?

Chapter 2

The Color Charge Maxwell Equations

§ 1. The Laws of Color Charges

The color charge Maxwell equations are known as "color pixel form" because each color equality is valid at every spatiotemporal point in space-time. These color charge equations are the color charge distribution law, the color charge gluonic wave law, the color charge monopole law, and the color charge current law.

➤ The color charge distribution law:

$$\nabla \cdot \left[\pm r\left(n_{\bar{r}}\bar{r} + n_{\bar{b}}\bar{b} + n_{\bar{g}}\bar{g}\right)\vec{a}_x \pm b\left(n_{\bar{r}}\bar{r} + n_{\bar{b}}\bar{b} + n_{\bar{g}}\bar{g}\right)\vec{a}_y \pm g\left(n_{\bar{r}}\bar{r} + n_{\bar{b}}\bar{b} + n_{\bar{g}}\bar{g}\right)\vec{a}_z \right] = \rho_V \quad (1.1)$$

$$\nabla \cdot \left[\pm \vec{a}_x \quad \pm \vec{a}_y \quad \pm \vec{a}_z \right] \cdot \begin{bmatrix} n_{\bar{r}} r\bar{r} & n_{\bar{b}} r\bar{b} & n_{\bar{g}} r\bar{g} \\ n_{\bar{r}} b\bar{r} & n_{\bar{b}} b\bar{b} & n_{\bar{g}} b\bar{g} \\ n_{\bar{r}} g\bar{r} & n_{\bar{b}} g\bar{b} & n_{\bar{g}} g\bar{g} \end{bmatrix} = \rho_V \quad (1.2)$$

$$\nabla \cdot \vec{q}_c = \rho_V \quad (1.3)$$

where "ρ_V" is the color charge volume density for an arbitrary two-dimensional color brane, "$n_{\bar{s}}$" is the coefficient for each color charge pair per square meter. Not all possible color charges may be present. The (1 × 3) matrix and the (3 × 3) matrix are the matrices of the color charge displacement field "$\vec{q}_c$" which is given in ±colorness/meter². The color flux density "ϕ_c" is given in (±colorness potential·second)/meter².

$$\boxed{\begin{array}{c} \vec{\phi}_c = \varsigma \vec{g} \\ \vec{I}_k = \xi \vec{q}_c \\ \vec{J} = \zeta \vec{g} \end{array}}$$

Figure 1. The Color Charge Medium Parameters.

where "ς" is the color charge permittivity in (±colorness/m)² / Newton / (potential·second), "ξ" is the velocity of the color charge permeability or the speed of light in a medium $"v=1/\sqrt{\mu/\varepsilon}=1/\sqrt{LC}"$ in (m/s), which is related to the ability of the medium to behave as an inductor "L" and a capacitor "C", and "ς" is the velocity of the color charge conductivity per unit of force in $(\pm\text{colorness/m})^2$ per (Newton·second). Moreover, the equation $"\partial \vec{j}/\partial \vec{s} = \partial \vec{g}/\partial \vec{t}"$ describes how the color charge source current varies with space equal to the way the gluonic field of color charges varies with time.

➢ The color charge gluonic wave law:

$$\nabla \times \varsigma \vec{g} = -\frac{1}{c}\frac{\partial \vec{\phi}_c}{\partial t} \qquad (1.4)$$

The gluonic field is given by "$\vec{g}$" in Newton/±colorness, and "ϕ_c" is the color flux density. The gluonic field may vary in a perpendicular direction to the color flux density. The colorness induced in a closed surface area is proportional to the rate of change of the color flux density that the color brane encloses. Every time the color charge gluonic field changes there is the creation of a color flux density. This law is a consequence of the total color charge energy conservation law. Every time there is a variation of the color flux density; a variation of energy in the medium, a color charge current is generated in order to keep the color flux density constant.

➢ The color charge monopole law:

$$\nabla \cdot \vec{\phi}_c \neq 0 \qquad (1.5)$$

A color string may have either a color charge displacement field or an anti-color charge displacement field. So, a single color string may be considered a color charge monopole. Color strings may form color charge dipoles. This law states the possibility of creating a color charge monopole. Thus, the total color flux density through a closed surface is not zero.

➢ The color charge current law:

$$\nabla \times \vec{I}_k = \vec{J} + \frac{\partial \vec{q}_c}{\partial t} \qquad (1.6)$$

where "J" is the color charge source current density ($\pm$colorness/second/meter^2) that describes the color charge distribution and the velocity of the color charges, and "$\vec{q}_c$" is the color charge displacement field ($\pm$colorness/meter^2), color charges flowing from their source to their sink. The color charge field strength "$\vec{I}_k$" is the ($\pm$colorness/second/meter). The color charge source current and the connection, completely determine all the physical properties of the system. The color charge source current may affect the curvature of the spatiotemporal medium of the connection.

The curl of "$\vec{I}_k$", or the "swirliness of "$\vec{I}_k$", is equal to the color charge source current density, the amount of color charge source current per unit area, plus any change in the color charge displacement field. The second term is often called the color charge displacement current since it has to do with the capacitive behavior of the medium.

Integral Color Pixel Form

$$\int_S \vec{q}_c \cdot d\vec{s} = \int_V \rho_V \, d\vec{v} = \pm k$$

$$\int_C \vec{sg} \cdot d\vec{l} = -\frac{1}{c} \int_S \frac{\partial \vec{\phi}_c}{\partial t} \cdot d\vec{s}$$

$$\int_S \left(\vec{\phi}_c \right) \cdot d\vec{s} \neq 0$$

$$\int_C \left(\vec{I}_k \right) \cdot d\vec{l} = \int_S \left(\vec{J} + \frac{\partial \vec{q}_c}{\partial t} \right) \cdot d\vec{s}$$

Figure 2. The Integral Color Charge Maxwell Equations.

§ 2. *What are the color charge maxwell equations in six-dimensional space-time?*

It is possible to reformulate some color charge Maxwell equations in color pixel form using the Tem Operator:

From previous research, the Tempus Operator, or Tem Operator, for three-dimensional time is denoted as

$$\odot = \vec{\Re}_\tau = -\frac{1}{c}\frac{\partial}{\partial t_x}\vec{a}_{t_x} - \frac{1}{c}\frac{\partial}{\partial t_y}\vec{a}_{t_y} - \frac{1}{c}\frac{\partial}{\partial t_z}\vec{a}_{t_z} \qquad (2.1)$$

The Color Charge Maxwell Equations in six-dimensional space-time are:

$$\begin{aligned} \nabla \cdot \vec{q}_c &= \rho_V \\ \nabla \times \varsigma \vec{g} &= \vec{\Re}_\tau \times \vec{\phi}_c \\ \nabla \cdot \vec{\phi}_c &\neq 0 \\ \nabla \times \vec{I}_k &= \vec{J} - \left(\vec{\Re}_\tau \times \vec{q}_c\right) \end{aligned} \qquad (2.2)$$

§ 3. What is the significance of the Color Charge Maxwell Equations?

The significance of the Color Charge Maxwell Equations is that they are the paramount tools to calculate effect over time in color charge field potentials. The equations have been designed using a contemporary mathematical style not the original style of James Clerk Maxwell Equations which were very long and complex but with greater substance related to the role of the dynamic spatiotemporal background. It was the practical electrician and eminent physicist Oliver Heaviside who express the equations in a pared down style that has made the equations easier to understand, but has also to the displeasure of some researchers that have felt the loss of data and the lack of further research in the reformulation of the original equations. It has been said that Maxwell provided the near-complete mathematical description of the behavior of electrical systems, but Heaviside simplified Maxwell's equations to the practical level of electrical engineering.

The causal link between induced color charges source current and the induced color charge displacement field appears simultaneously as a duality from the single causal source of the inertial momentum of the color charges. These two manifestation are a projection onto the spatial plane and the temporal plane from the spatiotemporal trajectory of the actual color charges as the movement is described mathematically with respect to space or time. The color charges field equations are described as time-dependent color charge potentials and rely on the spatiotemporal background for their motion and medium of propagation. Space-time emerges as dynamic waves and as the domain of physical events.

§ 4. What are the Color Charge Maxwell Equations?

The Color Charge Maxwell Equations describe simplified ways for the quantum natures of the color charge displacement field and the gluonic field in a Planck medium. Hence, the color charge displacement field and the gluonic field are not independent but instead they are conjugate. The two fields are the two faces of the same coin, and the mint is color string theory.

The Color Charge Maxwell Equations (CCMEs) are a mathematical summary of the quantum gravity theory named "A Dynamic Theory of Space-Time: A Matter of Waves." They describe how both the color charge displacement field and the gluonic field arise from color charges and currents, how the color charges propagate and influence each other. These quantum equations along with the strong nuclear force quantify most of the quantum process that gluons experience, including quarks and anti-quarks, hadrons, and elementary particles and anti-particles. They are the foundation of quantum physics at the Planck scale.

These CCMEs are a compact set of equations that completely specify the color charge displacement field and the gluonic field with the tools of vector calculus and quantum field theory. The color charge field equations may be derived from the divergence and the curl of the color charge field. The CCMEs provide the divergence and the curl of the color charge displacement field and the gluonic field in terms of the color charges and their currents. Hence, the color charge field and the gluonic field may be calculated from the displacement and overall distribution of the color charges for any set of circumstances.

The CCMEs build upon "A Synthesis of Quantum Gravity" theories and equations to provide an incisive realization into the quantum origin of classical theories like electromagnetism and gravitation. It is possible to theorize that a color charge displacement field and a gluonic field may exist in the presence of color charges, or in the absence of gluons. The color charge field is an oscillation, or a traveling wave, moving at $"v = 1/\sqrt{\mu/\varepsilon} = 1/\sqrt{LC}"$ meters/second. Does a photon, the carrier of the electromagnetic force, also consist of color charges? If affirmative, would it not exhibit the behavior of color charges?

If light emerges from the fields and currents of color charges, the displacement of color charges and the emergence of gluonic fields can manifest the gluons and quarks that bind to make photons which are light quanta and waves. From those classical waves emerges the whole electromagnetic spectrum. Would a color charge technology create the future devices that could become the light sources, lasers, EM waves, for the quantum communication devices of tomorrow? The future industries of the world and the new emerging technologies may be based upon these concepts, to improve the quality of life of people around the world and the productivity of workers, businesses, and governments.

PART II

COLOR SUPERSYMMETRY

Chapter 3

Color String Supersymmetry

§ 1. The Theory of Color Supersymmetry (COSUSY)

The symmetries found in physical reality provide a crucial framework to our understanding of our universe, from the four fundamental forces of nature to the unification of all forces that existed at very high energies in the early universe according to our present understanding.

Since the 1970s, supersymmetry was proposed as a potential symmetry that could unify all types of particles in our physical reality, from fermions to bosons and other particles in between. The supersymmetry connection depends on the peculiar property of particles called spin, and imaginably has the potential to uncover a new way to understand the laws of physics that rule our universe. Hence, it is important to expand on the Theory of Color Supersymmetry or *COSUSY,* to explain and clarify the potential of color supersymmetry within the context of color charges.

§ 2. The Potential of Color Symmetries

Historically, physicists and mathematicians have used the potential of symmetries to uncover the elemental relationships and fundamental connections that exist in the physical reality of our universe. During the fourth century BC the Greek philosopher Aristotle believed that objects tend toward a point due to their inner gravitas, or their heaviness. As Aristotle thought about gravity in a way that compared very heavy objects to lighter ones, he was discovering a symmetry of nature. Gravity is a universal law of nature. This insight from Aristotle led other scientists in history to leap forward to a better and more comprehensive understanding of the role that gravity plays in nature.

At the present time, quantum physics researchers are bemused over the peculiar properties of color charges and their implications. What causes color charge current to flow? How could gluons that spin make color charge currents to flow through quantum conductive mediums? What is the driving force for circulating color charge currents? Is light made of gluons? Is a gluon a wave and a particle?

All these questions beg for an answer that has built up to a Quantum Theory of Color Strings which provides an effective mathematical proposal to unify these distinct branches of a quest under a simplified set of Color Charge Maxwell Equations. It is no coincidence that most of the realizations and theories of science are built upon the foundation made by predecessors as each new researcher takes a step further or a great leap forward.

It is possible to theorize that all physical laws involving color charges should be the same regardless of your position or momentum. Hence, it is important to also commend Color String Theory with the notion of space-time as the source of space, time, and color, to conserve the color supersymmetry of physical reality in nature. Moreover, the effect of the color graviton, as a quantum manifestation of gravity, leads us to a more profound understanding of the color force. The color conservation laws, not limited to the conservation of color charge momentum, and the conservation of colorness energy, depend on color supersymmetry. The law of conservation of colorness energy support the constancy of the laws of color charges. The same experiment on color charges done repetitively at different locations and times should yield the same results. So, there would be a color supersymmetry across space, time, color, momentum, and energy.

§ 3. The Theoretical Predictions of Color Supersymmetry

Color Supersymmetry predicts a color charge partner for each color charge in the Gluon Standard Model, to explain why a color charge has the property of colorness. From previous research, it is proposed that in the absence of other fields, the refractive color force field, or the refractive gluonic field, is a consequence of a virtual color charge symmetrical partner, or an equivalent virtual color charge, to every color brane existing in homogeneous and isotropic space-time. The boundary between the color brane and space-time acts as a mirror due to space-time distortion, or a reflector, manifesting the image of an equal and opposing virtual color symmetrical partner to the existing color charge in space-time.

The only color force field lines from the color charge that cross the boundary between the color charge and space-time are at ninety degrees to the surface of the boundary. Those orthogonal color force field lines extend radially outward in space-time, while all other non-perpendicular color force field lines are reflected back toward the color charge, manifesting a virtual color symmetrical partner, on the space-time boundary mirror. This exerts a refractive color force field from free space-time on a color charge pixel "q_c". Thus, the color charge permittivity of free space-time is the effect of the refractive color force of the color charge, at a distance from its center of colorness.

The Gluon Standard Model will predict what future experiments may show about the building blocks of quantum color objects. The current Standard Model is beginning to be considered incomplete. Color Supersymmetry, or COSUSY, is an extension of the Gluon Standard Model that focuses on bridging some of the gaps in the mathematical framework and the spatiotemporal dimensional assumptions. The Gluon Standard Model predicts a color charge partner for each color brane or gluon. These virtual color charge partners may resolve considerable problems regarding the mass of the newly found Higgs boson. However, no experiment is likely to show the virtual supersymmetric color charge partners unless a technology that can detect the reflective spatiotemporal distortion, or the reflection of color force fields by a reflective spatiotemporal boundary, is used. But history teaches us to expect the unexpected when it comes to the development of experimental equipment.

The Gluon Standard Model predicts that elementary particles should not be massless, a prediction that agrees with physical observation. The Higgs boson has been theorized as a mechanism, but not the only mechanism, that endows elementary particles with the property of mass. Nonetheless, why is the Higgs boson lighter than expected? It was expected to be a heavier boson due to its interactions with elementary particles. Color Supersymmetry predicts a virtual supersymmetric color charge partner that would offset the contribution to the mass of the Higgs boson by any elementary particle. This prediction makes the existence of a lighter Higgs boson possible.

The virtual color charge partner concept may resolve the hierarchy problem which is the large discrepancy between the aspects of the weak nuclear force and gravitation. So far, there is no scientific consensus on why the weak nuclear force is 10^{24} times stronger than gravity.

A COSUSY extension to the Gluon Standard Model would resolve the hierarchy problem within gauge theory, by ensuring that quadratic divergences of all orders will nullify in perturbation theory. Color Supersymmetry is also stimulated by several solutions to theoretical problems, to ensure a reasonable behavior at very high energies, and to provide many desirable mathematical properties in a general fashion.

The virtual supersymmetric color charge partners would interact with the same color forces and fields as the real color charges but they would have imaginary charges. The strong nuclear force of virtual supersymmetric color charge partners would have the same strength as color charges during the high energy conditions of the early universe.

The same prediction applies to the weak nuclear force, and the electromagnetic force. A Dynamic Theory of Space-Time is a Quantum Gravity Theory that predicts a unification of all electroweak forces, the strong nuclear force, and gravitation.

Gluons	Red (r or ir)	Blue (b or ib)	Green (g or ig)	Antired ($\bar{r}$ or $i\bar{r}$)	Antiblue ($\bar{b}$ or $i\bar{b}$)	Antigreen ($\bar{g}$ or $i\bar{g}$)
Real Color	$\pm \frac{2}{3}e$	$\pm \frac{2}{3}e$	$\pm \frac{2}{3}e$	—	—	—
Imaginary Color	$\pm i\frac{2}{3}e$	$\pm i\frac{2}{3}e$	$\pm i\frac{2}{3}e$	—	—	—
Real Anticolor	—	—	—	$\mp \frac{e}{3}$	$\mp \frac{e}{3}$	$\mp \frac{e}{3}$
Imaginary Anticolor	—	—	—	$\mp i\frac{e}{3}$	$\mp i\frac{e}{3}$	$\mp i\frac{e}{3}$

Figure 1. The Chromodynamic Color Charge to Electrical Charge Correspondence of Gluons.

Imaginary Color Charges	$i\bar{r}$	$i\bar{b}$	$i\bar{g}$
ir	$-r\bar{r}$	$-r\bar{b}$	$-r\bar{g}$
ib	$-b\bar{r}$	$-b\bar{b}$	$-b\bar{g}$
ig	$-g\bar{r}$	$-g\bar{b}$	$-g\bar{g}$

Figure 2. The Imaginary Color Charges of the COSUSY Model.

Color Supersymmetry connects elementary particles like fermions and bosons through the property of spin. Bosons have a spin of 0, 1, or 2. Fermions have a spin of ½. COSUSY predicts that each virtual supersymmetric color charge partner to a color charge with a spin would differ by ½ of spin. Bosons and fermions differ in spin and other behavioral properties. Bosons tend to be in the same quantum state and flock together like birds of a feather, while fermions prefer to be in a different quantum state and separate. Nevertheless, COSUSY draws the bosons and fermions together.

It is possible that the existence of lighter virtual supersymmetric color charge partners may also account for the missing baryonic matter or energy of the observable universe under the current Standard Model. These lighter virtual supersymmetric color charge partners are stable, colorless, and interact weakly with the elementary particles. From previous research, it was theorized that color pixels varied in their intensity of their colorness, through the processes of chromomeiosis or chromosynthesis of color string theory.

From previous research, any two fractional values of colorness of two subpixels, or two sub-antipixels, may add, up to the next upper value of colorness, to a superpixel, or super-antipixel, of stronger color. Any pixel, or antipixel, may divide, down to the next lower value of colorness, into two subpixels, or two sub-antipixels, of the same, but weaker color, where each subpixel, or sub-antipixel, has a value of colorness that is one half the value of the superpixel, or super-antipixel. Thus, subpixels, or sub-antipixels, have a weaker color charge than the color charge of the original superpixel, or super-antipixel. Individual color charge of pixels may not be conserved, but the collective color charge of all pixels, or antipixels, is conserved.

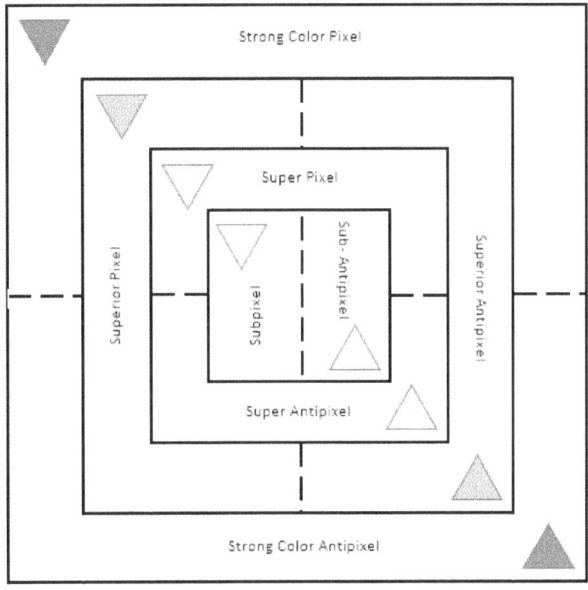

Figure 3. An Illustration of Chromomeiosis or Chromosynthesis.

Therefore, COSUSY provides the framework that builds upon the foundation of the Gluon Standard Model to create a more comprehensive explanation of the physical reality of our universe. Regretfully, the current Standard Model provides an incomplete picture of the observable universe to say the least. On the other hand, there is more to discover through theoretical or experimental research.

§ 4. A Concerto of Color Charges: Who is playing the color strings?

Since the 1970s, researchers began to play the strings of a concerto without colorness, looking carefully into the quantum properties of particles, specifically into the property of quantum spin to find an area or instance of symmetry in our physical reality. If symmetry was found, then there would be reason to have common ground between the whole-spins and the half-spins, or between the bosons and fermions, to unite these very different quantum particles. So, the quest for supersymmetry started to get the attention of the particle physics community of researchers. At the quantum level, the elementary particles in our physical reality have the peculiar property of spin. This property was experimentally discovered in particle accelerators to deflect the trajectories of elementary particles through a magnetic field, in the same way as a classical spherical mass would deflect, if the spinning metallic mass were electrically charged.

However, elementary particles do not spin like a classical metallic mass, but the behavior of the quantum particles resemble that of the classical objects in specific experiments. At the quantum level, elementary particles can only have a specific amount of spin. Each specific quantum particle has a unique spin.

For example, a photon has a spin of 1, while an electron has a spin of ½. It is possible to describe spin as a tumble or tumbling motion through space-time. A tumble that results from a quantum particle spinning on each of the three spatial planes of a Cartesian coordinate system for a specific fraction or whole number of times, to return to the specific original orientation, or state, that it previously had in space-time. Quantum particles may or may not share the same specific amount of a quantum property like spin, charge, or mass.

Hence, there are quantum particles that have whole-integer spin, 0, 1, or 2, and there are other quantum particles with half-integer spin, 1/2, 3/2, or 5/2. The whole-spins are bosons or carriers of the force like gluons, photons, $W^{\pm}$, and Z^0 particles. The half-spins are the fermions like quarks, electrons, nucleons, and neutrinos, etc., that are the constituents of physical reality.

§ 5. The Spinning Reflection of a Color Charge

Is the spinning virtual supersymmetric color charge partner a spinning mirror?

In Color Supersymmetry, every fermion would an imaginary partner, or an i-partner for short, that is a supersymmetrical particle in the fermion domain, and similarly, a boson has an i-partner in the boson domain, with the same charge and mass, but a different spin. Nevertheless, the spin of an i-partner is an opposite spin that returns the i-partner to its original i-state. So, imaginary color supersymmetrical partners exist in their i-partner domains.

Therefore, the symmetry would hold in the physical reality of our universe, symmetry is not broken. The Color Supersymmetry is achieved in a way that it predicts the masses, charge, and spin of all fermions and bosons correctly. The evidence of the COSUSY is found everywhere and at any time on the spinning mirror around every elementary particle and every color charge. Thus, COSUSY is found in physical reality in the spatiotemporal boundaries where the pressure of gravitation forms lenses or dense regions of gravitational mirrors on boundary regions of high energy density, mass, or matter. It is possible that fine lasers may be used on the boundary regions to verify the imaginary or reflective property of Color Supersymmetry.

Once in a great while, there is a concept or an innovative idea in quantum physics that holds great promise for the unification of the four forces of nature. If a unique idea solves a large number of mysteries of nature in a single swift action to predict the outcomes of physical processes or experiments, that unique idea or theory would cause a significant impact on science and encourage the interests of physics researchers everywhere. If its predictions are true and accurate it can start a revolution in science with a paradigm shift in the understanding of our universe. This was exactly what happened with Newton's Principia, or Einstein's General Theory of Relativity. At the present time, it is a great mystery of science why the elementary particles of the current Standard Model have such small masses compared to the Planck mass, or why present theories of the four forces of nature resist unification under well accepted theories, or where is the missing baryonic mass of our universe. Fortunately, COSUSY and a new theory of quantum gravity may provide a solution to each of these riddles, while predicting a large number of color supersymmetric partner particles and other new particles. Virtual or imaginary particles may not be physically manifested, so new experimental techniques may be used to indicate why and how they impact their physical reality.

The incentive for COSUSY has existed since the theory of quantum chromodynamics was introduced in the 1960s by eminent physicists Murray Gell-Mann and Yuval Ne'eman, and developed further by other researchers. Color Supersymmetry acknowledges and resolves a problem from the early days of quantum mechanics with the charge of an electron and its diminutive physical size as a point particle. A charged particle has its own inherent voltage potential and an electric field. Hence, the inherent energy of an electron demands a larger physical size according to current laws of particle physics. The higher the energy the lesser the physical size should be; so, an electron that is really a point-like particle should have an incredible amount of energy. However, we know from experimental evidence and precise calculations that is not the case.

Hence, the size of the electron should be approximately $5 \times 10^{-15} m$, or larger than the size of a nucleon, and its inherent energy is a finite amount. The existence of colors and anti-colors from COSUSY that make the virtual particles, and photons of free space, predicts that the electron is also made of a colorless combination of color charges with an electrical charge due to parity. This prediction also includes electron-positron pairs, or colorless combinations of color-anti-color pairs.

During a high energy collision between two photons, an electron may be produced, while an electron may also produce a photon. COSUSY predicts the same color constituents in either an electron or a photon. Both particles consist of color charges and their properties. It is also interesting to consider the annihilation of an electron with a positron because only the fluctuation of the electron remains, resulting in the diminutive size of the electron and its considerable electric charge.

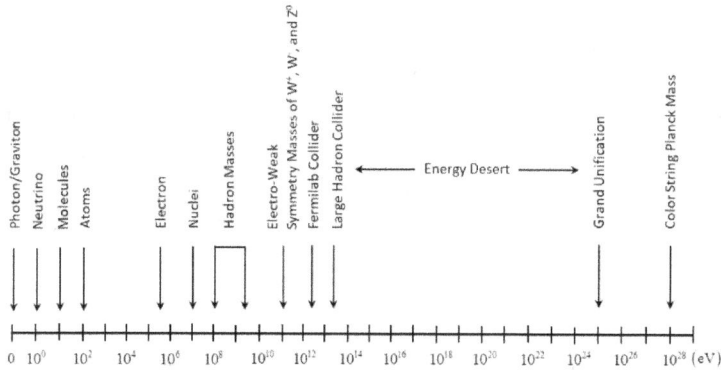

Figure 4. The Mass-Energy Scale.

$$Length-Scale = \frac{hc}{E} \approx \frac{10^{-4}}{Energy\ (eV)}(cm) \approx c \times Time-Scale \quad (5.1)$$

where "h" is the Planck constant, "E" is energy, "c" is the speed of light, and time-scale is usually given in seconds.

The tension in a color string would be on the order of the Planck force about 10^{44} Newtons, which corresponds to a mass of about 10^{28} eV. Thus, only the particle with a massless state in color string theory would be consistent with physical reality. Actually, the observed mass of all the elementary particles would be negligible if compared to the Planck mass of 10^{28} eV. The above figure shows the wide gap between the mass of known particles and the Planck mass. It is interesting to point out that there is an extensive region that is labeled an energy desert. The energy desert, a theorized gap in energy scales, has been considered a vast energy domain where nothing of interest happens until the Planck energy of about 10^{24} eV is reached. The mass or energy may be denoted as binding energy for some systems of mass such as nuclei, atoms, and molecules.

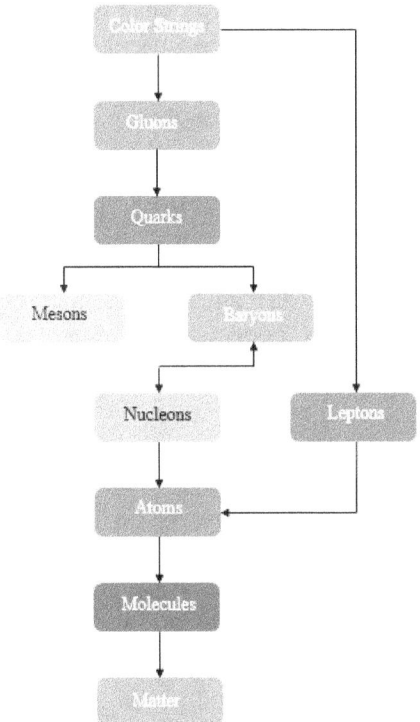

Figure 5. The Structure of Matter.

So, what does the annihilation of particles made of color-anticolor charges have to do with COSUSY?

The quantum annihilation of particles that consist of color-anticolor pairs occurs because the there is a symmetry in the Theory of Color Supersymmetry between color and anticolor that safeguards the properties of a particle, allowing the particle to have its size, charge, mass, and color properties.

The main idea of COSUSY revolves around the presence of an additional symmetry between bosons and fermions that safeguards the properties of colorness and endows the particle color charge or mass to be diminutive size and smaller energy compared to the Planck scale.

COSUSY allows a color charge partner for every elementary particle. This adaptive COSUSY concept would double the number of known elementary particles, as an imaginary partner exists for every boson and fermion with opposite spin and similar mass and charge, or for any newly found particle.

Consequently, the color supersymmetry normalizes the particle size and mass to the observed values. The lower values of COSUSY may be expected to be measured at lower energy level that the theoretical grand unification, to allow the coupling constants of the electroweak and strong nuclear forces to be unified at the theoretical grand unification scale, and would create stable and neutral COSUSY particles that would explain the missing baryonic mass and energy in our universe.

As the coupling constants are viewed on a log-log scale as a function of energy, they do not seem to meet as shown below on the left. (de Boer, 1994) However, if i-partners are added as predicted by color supersymmetry, the constant would meet at the color forces unification scale of about $10^{25} eV$.

Aside, the Large Electron-Positron (LEP) collider is regarded as the largest electron-positron accelerator ever built. (CERN, 2001)

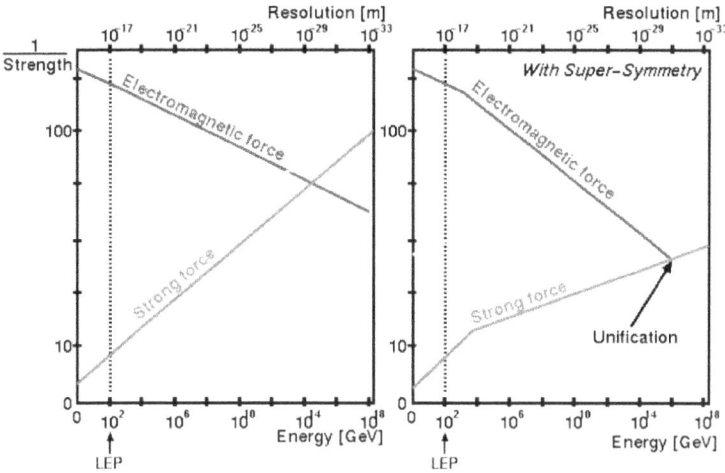

Figure 6. The Strength Reciprocal versus The Energy for the Color Forces of Nature.

At the present time, the predicted values of mass and other physical properties of the known particles in the current Standard Model from Plank units and fundamental principles are seventeen orders of magnitude larger than the observed values of the heaviest elementary particles. Something is afoot! The Higgs boson and other mechanisms for gaining mass should be near or close to the expected value of the Planck mass value. COSUSY provides a theory that addresses these challenging irregularities and the hierarchy problem.

There is also the present uncomfortable verity that the strong nuclear force may not unify with the electroweak force because of the lack of experimental evidence so far. The three projected lines of the EM-Weak Nuclear-Strong Nuclear Forces do not intersect at the expected unification point. What is missing in the present theory?

As effective experiments will be hopefully conducted, the existence of the i-partners for color charges and elementary particles should provide the evidence of the heaviest mass for each i-partner of the known elementary particles and color charges of the Gluon Standard Model.

The Large Hadron Collider was designed to provide answers to these questions but it seems that a different type of experimentation is needed.

This type of experimentation was a major purpose for designing and building the LHC. However, the expected particles have not been found so far, and the limits of the magnitude of the calculated mass on the particles have increased tremendously, that researchers are still looking for the solution to the hierarchy problem, So, COSUSY is theorized to provide a solution to this problem and others. COSUSY provides theoretical solutions to why the color supersymmetrical i-partners are so large and why the masses of elementary particles are so diminutive. The incentive of COSUSY as an enchanting and beautiful theory of color supersymmetry is also a pragmatic inducement since other theoretical approaches have fell short.

It is very important to predict using the color supersymmetry theory, but also to test those ideas theoretically and empirically to achieve the graceful and potent results that accurately describe our physical reality. Thus, a great deal of theoretical and empirical work remains to be done. Experimentation and replicates of results should provide veracity and factual evidence. COSUSY is in the framework of color string theory, not as a means to an end, but as a substructure that upholds the theory from the Planck scale to higher structures of the macroscale of all physical objects of space-time, mass, energy, and charge.

One simple truth is that a researcher or experimentalist that is also seeker of physical truth in nature will not find what is not there to begin with, but the real question becomes what other phenomenon in nature, that is present, could produce the same effects. If the quest is not continued, how would nature reveal to the researcher or experimentalist its mysteries? To find or not find what is sought is in itself an achievement in science. After all the evidence and facts are proposed, nature will always be the final judge of the veracity of a theory or an idea.

What is the current status of Color Supersymmetry?

The Gluon Standard Model of elementary particles may be very successful and complete to enhance the present Standard Model that is regarded as incomplete, even though it has served an interim role for successful predictions of the properties of particles and the features of physical reality. As science approaches the ultimate truth of our universe, ideas may have to be replaced by greater ideas, and theories by more profound theories. So, this is the way of the advancement of science.

Color supersymmetry is theorized to fill the gap by doubling the number of known elementary particles in the current Standard Model. This generalization of symmetries is relativistic in nature because it involves spatiotemporal reflection about the particle mass or energy due to the known properties of space-time. This effect creates a supersymmetrical i-partner for every particle, and virtual mass. COSUSY predicts a menagerie of i-partner particles with masses that include the masses of to the Z^0 and $W^\pm$ bosons, but not from a very high-energy collision. Hence, instead of a violent collision of particles, the quest may require a subtle search around their spatiotemporal boundaries.

The supersymmetrical particles had been expected all around other particles anywhere, with masses that were not heavier than the masses of particles known to be in the quantum regions that were previously researched, without ever finding indirect effects in processes that involved low energies. These supersymmetrical particles were not detected in the Large Electron-Positron Collider in the 1990s, which initially cast doubt about their existence. (Arkani-Hamed et alia, 2004)

Hence, it is becoming increasingly hopeful that Color Supersymmetry may include the original features sought after by physics researchers, namely, a unification of forces, a particle for the missing fermionic mass, and an explanation for the mass of the Higgs boson. A color string loop around an arbitrary spatiotemporal point where the two ends of the color string may join together to manifest a color graviton. Therefore, is it possible that the existence of color gravitons may endow the Higgs field? The idea of Color Supersymmetry has the potential to provide the desired features.

Reassessing the ideas and premises of Color String Theory.

Let us theorize that Color Supersymmetry exists in our universe. If two particles are entangled, would the entanglement also exist between the i-partners of the same particles? So that, if the color supersymmetrical particles are separated there would be a spatiotemporal quantum conduit between them that would serve as a tachyonic circuit. Regardless of the spatial distance, the supersymmetrical particles may be entangled through time to exchange quantum states instantly through the spatiotemporal medium.

Moreover, supersymmetrical i-partners exist but cannot be detected by collisions designed for physical particles and their trajectories, not virtual particles. The i-partners would not be detected by a certain high level of energy outside the sphere of influence of the i-partner that is inside the spatiotemporal boundary of the corresponding particle. No exotic signatures of physical particles are expected, only the signatures of virtual i-partners with significant virtual limits. No need for ghost particles that disappear after creation so they do not travel away to be detected.

Therefore, the supersymmetrical i-partners are long-lasting and very spatially adjacent to the corresponding particle which may not stay at a fixed virtual distance due to quantum field variations and spatiotemporal motion that affects the reflection of the virtual i-partner. The approach of an optical search for supersymmetrical i-partners of low-energy Standard Model particles would be useful for Color Supersymmetry. The unrelated background of low-energy particles that may exist in the medium of detection, so an experimental technique to filter out the effects of these unrelated low-energy particles that may indicate Color Supersymmetry would be necessary.

What are current researchers searching for? It is the quest to discover and understand the mysteries of nature that drives researchers of supersymmetry. COSUSY has the potential to provide all the features that researchers hoped and searched for. Thus, more research is needed to validate its veracity and the substructure of its elements.

Summarizing the potential benefits of Color Supersymmetry as a theoretical framework:

- COSUSY would stabilize the hierarchy of mass scales that are found in particle physics, such as the Planck mass, the electroweak force, and the grand unification scales.

- A great candidate for the missing fermionic mass in our universe in the form of the lightest color supersymmetric particle. A virtual particle that will not have to be theorized heavier than originally thought, and still be consistent with the experimental mass of the Higgs boson.

- The very precise measurement by the Large Electron–Positron Collider of the electroweak mixing angle which would agree perfectly with color supersymmetry as a unified theory of color charges. The electroweak mixing angle is a fundamental parameter of the Gluon Standard Model that quantifies the relative strengths of the weak force and electromagnetism, and governs how a Z^0 boson couples to a fermion.

- The theoretical existence of the Color Graviton may endow a Higgs field. The indication that the calculated mass of 125 GeV/c^2 for the lightest supersymmetric Higgs boson agrees with the experimental value. The coupling of the Higgs to other particles would be very close to the Gluon Standard Model, which is consistent with what has been measured as yet.

- COSUSY would stabilize the electroweak vacuum.

The potential of COSUSY as a theoretical framework has the incentive that it can stabilize the hierarchy of mass scales such as the Planck scale, the electroweak scale, and the grand unification scale. COSUSY indicates it can also represent a contender for the finest color supersymmetric particles that may constitute the missing fermionic matter of the universe. After the very precise measurement of the Large Electron–Positron Collider of the electroweak mixing angle, it is possible to realize that COSUSY is in exact agreement with the grand unification theories. The calculated mass of the finest color supersymmetric Higgs boson, with an empirical mass of approximately 125.1 GeV/c^2 determined by the Large Electron–Positron Collider, agrees with the probable calculated value.

Hence, there were very high expectations and enthusiasm during the initial operation of the Large Hadron Collider about supersymmetric particles, but not about color supersymmetric i-partners because the i-partners are virtual particles that act as if they were physically there from the perspective of the real partners. Even with COSUSY in the underlying theory, the lack of physical detection does not nullify the underlying principles of the theory or its potential success. Nor the supersymmetric particles have to be necessarily imagined heavier than originally thought by researchers. Since i-partners are not necessarily heavier, there is no need to do much more fine tuning to solve the electroweak problem. The empirical predictions of the Large Hadron Collider about the mass of the Higgs reinforces the stability that COSUSY provides to the electroweak vacuum, while the Higgs couplings to other particles are more consonant with current measurements.

The increase in the GeV scale of the Large Hadron Collider to a TeV scale may bring the unexpected discovery of the veracity of COSUSY from the theoretical realm to the existential effect of the underlying principle of the theory. The outcome is likely to be new physics, more precise predictions and measurements for the electroweak area, and a greater clarification of the missing fermionic matter. A higher-energy Large Hadron Collider should indicate the existence of any weakly-interacting massive particles or their non-existence. As science develops further, COSUSY may allow us to develop a deeper understanding of the six-dimensionality of space-time and how that affect the missing fermionic mass of the universe, providing a much greater insight into physical reality of particles and waves and their spatiotemporal background as well as the face and properties of COSUSY in topological condensed-matter systems.

COSUSY may be described as a symmetry in nature that may not be really there but would have virtual implications in theories of quantum gravity under the gluon standard model. COSUSY could play an important role as researchers explore new physics below the Planck scale of physical reality and virtuality. The interactions of color strings will provide greater understanding at the quantum level as the substructure of matter, energy, and space-time.

There is no need to modify the early expansion evolution of the universe when the masses of the COSUSY i-partners can provide the missing fermionic mass or solve the hierarchy problem. COSUSY has the potential to play a crucial role in new particle physics because the i-partners could get rid of several infinite quantities that may otherwise emerge from the calculations of high-energy particle interactions for the unification of the forces of nature. That is not to say that more research is not needed for the early history of the universe or whether the universe had or did not have a regular smooth adiabatic expansion as our understanding grows beyond current assumptions of initial conditions. COSUSY entails the duplication of the number of the known elementary particles. For instance, a boson or a fermion would have a COSUSY i-partner where the fermionic i-partner would have its bosonic COSUSY i-partner. So, gauge bosons such as a gluon, a photon, $W^{\pm}$, and Z^0, have their i-partners. These COSUSY particles exist as virtual i-partners to real fermions and gauge bosons. The i-partner virtual particles are the fermionic particles or gauge bosons into which the real fermions and gauge bosons are transformed by COSUSY. Color Supersymmetry may provide a better understanding of the partnership of fermions and bosons, and the couplings of these particles.

COSUSY is a complex mathematical framework based upon the theory of group transformations and a symmetry between bosons, which are particles with integer values of spin, and fermions, which are particles with half-integer values of spin or intrinsic angular momentum. The COSUSY concept is a key feature for the realization of the color graviton, a coherent quantum theory. COSUSY exemplifies local symmetry with spatiotemporal transformations that lead to the realization of the color graviton, a spin-2 boson with the source of gravitation as the stress–energy–momentum tensor, or the carrier of the gravitational force. In that local framework, COSUSY incorporates a color supergravity theory.

Symmetry refers to a property of harmonious and unchanging proportion, orientation, and balance. Mathematical symmetry is a more exact definition, to refer to an object that is unchanging under some transformation operation, including translation, rotation, scaling, or reflection. A triangle, that is rotated about its center through 120, 240, or 360 degrees, has three-fold symmetry by which the triangle appears to be in the same orientation. The three rotations bring the triangle back to its original orientation. The laws of physical reality manifest the symmetry of space and time through the conservation of momentum and the conservation of energy. Color Supersymmetry endows a particle such as a fermion to be transformed into a particle like a boson, without varying the characteristics of the particle carrier interaction or the underlying fermionic particle theory. The transformation of a fermionic particle into a gauge boson and its reverse transformation into a fermion involves a translation through the spatiotemporal medium under the relativistic effects of the General Theory of Relativity. Thus, COSUSY provides a correspondence between spatiotemporal transformations to the intrinsic angular momentum or spin of elementary particles. Hence, COSUSY allows a correspondence between the fundamental particles of mass, which include fermions and bosons, effectively reducing the number of particles to only the elementary particles of the substructure such as the gluons according to current understanding.

The theoretical paradigms of new physics rest on the General Theory of Relativity, Quantum Mechanics and the Gluon Standard Model, which have been discussed in previous research as being nested theories, or arranged in a hierarchical structure that is embedded in physical reality. (Nieves, 2020 and 2021) This hierarchical structure indicates the direction of experimentation toward a greater realization and the expansion of knowledge and advanced technologies.

Chapter 4

The Spatiotemporal Unitary Color Group of Order "n" for Color Charges

§ 1. Do Symmetries, Translations, Scaling, Reflections, and Rotations lead to the Discovery of the Laws of Nature?

Why is the Spatiotemporal Unitary Group important in physics? In the mathematics of Group Theory, the spatial (or special) unitary group of degree "n", denoted SU(n), is the Lie group of n × n unitary matrices with determinant of value 1.

The more general unitary matrices may have complex determinants with absolute value of 1, rather than real value of 1 in the spatial case. The temporal unitary group of degree "n" is denoted TU(n), is the Lie group of n × n unitary matrices with determinant with an absolute value of 1. The group operation is matrix multiplication.

The unitary groups would seem very natural if you are familiar with normalized complex vector spaces, like the Hilbert spaces of Quantum Mechanics. They are the length-preserving maps or the orthogonal groups of the complex realm. Particularly, U(n) is the set of all functions on C^n that preserve the normalized $u^\dagger u$ for all $u \in C^n$, where the symbol $\in$ indicates set membership and means u is an element of the set C^n, and $u^\dagger$ is the Hermitian conjugate of u, so that if $u = u^\dagger$, then $uu^\dagger = uu = u^\dagger u$.

Thus, a square matrix is Hermitian if and only if it is unitarily diagonalizable, or similar to a diagonal matrix, with real eigenvalues, which are the scalar factors, or changes, of nonzero eigenvectors, after linear transformations are applied to each of them. In linear algebra, an eigenvector changes at most by a scalar factor when a linear transformation is applied to it.

Aside, $\langle,\rangle$ is an inner product of C^n, called the standard inner product on C^n. The vector space C^n with the standard inner product is called the complex Euclidean n-space. Also, if the conjugate transpose of a square matrix is equal to its inverse, then it is a unitary matrix.

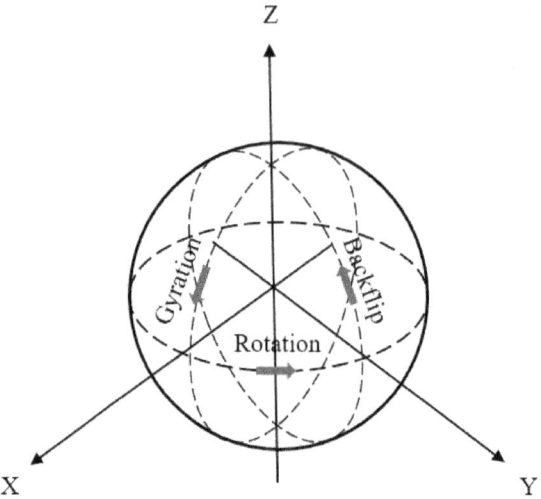

Figure 1. An Illustration of the Potential Components of Spin.

(spin → gyration (pitch) + rotation (yaw) + backflip (roll) → tumble)

It is important to reiterate that the spatial (special) unitary group of degree "n", denoted SU(n), is the Lie group of n × n unitary matrices with determinants of a value of 1. Other unitary matrices, that are more general, may have complex determinants with absolute value of 1, rather than with real value of 1 in the spatial (special) case. The group operation is matrix multiplication.

In Group Theory, SU(2) is also identical to one of the symmetry groups of spinors, Spatial Spin(3), that enables a spinor presentation of rotations.

The spatial (special) unitary group SU(3) is a subgroup of the unitary group U(3), consisting of all n × n spatial unitary matrices. The temporal unitary group TU(3) is a subgroup of the unitary group U(6), consisting of all temporal n × n unitary matrices. The symmetry group SU(3) appears significantly in the physics of elementary particles. Each of these symmetries refers to an underlying threefold symmetry in the physics of the strong interaction. So, SU(3) is the group of special unitary 3 × 3 matrices U.

Time emerges from space, then space endows more time. It is possible to regard a spatial distance "r" as equivalent to a temporal distance times the speed of light, "ct". From previous research, we may also regard SU(6) as representing a (3 + 3) formalism for six-dimensional space-time in terms of a combination of three spatial dimensions and three pseudo spatial dimensions. SU(6) corresponds to a general special unitary transformation on complex, both spatial and temporal, unitary vectors. The natural representation is that of (6 × 6) matrices acting on complex six-dimensional vectors. There are (9) parameters, ($n^2 - 1$) possible generators, from which only (9) generators, $(X_0, X_1, ... X_8)$ are applicable, and (8) generators, $(X_1, ... X_8)$, are equivalent to the SU(3) Gell-Mann generators. (Nieves, 2020)

The SU(n) groups find wide application in the Gluon Standard Model of particle physics, particularly the simplest case, SU(1), in the spatiotemporal waveform interaction at an arbitrary point, SU(2) in the electroweak interaction, and SU(3) in quantum chromodynamics. The TU(n) groups find wide applications in the Gluon Standard Model, especially in the gravitational field about a mass or a body of mass. From previous research, the gradient of the gravitational field is the gradient of the temporal scalar field near a massive object.

The SU(1) group is the trivial group or zero group, having only the subgroup consisting only of the identity element. The single element of the trivial group is the identity element that is usually denoted as such: 0, 1, or e depending on the context. If the group operation is denoted "·" then it is defined by $e \cdot e = e$. The trivial group is "cyclic" of order 1, and as such it may be denoted Z_1 or C_1. If the group operation is addition, the trivial group is usually denoted by "0". If the group operation is multiplication, then, the trivial group is typically denoted "1". Consequently, these definitions may be combined for the trivial or zero ring in which the addition and multiplication operations are identical and as a result, $0 \equiv 1$. The trivial group serves as the zero object in the category of groups, meaning that it is both an initial object and a terminal object. Hence, the trivial group, SU(1), finds wide application in the spatiotemporal waveform interaction at an arbitrary point of the medium.

As the retarded and advanced spatiotemporal wavelets interfere constructively or destructively at an arbitrary point, the spatiotemporal medium may expand, contract, or stay the same. All spatiotemporal wavelets from all directions add at a point denoted by "0", or extend the amplitude of the wavelet denoted as "1", where the subtraction property may be considered as a special case of the addition property in the interference of spatiotemporal waves. (Nieves, 2020)

As a compact classical group, U(n) is the group that preserves the standard inner product in the field of real numbers (R^n) and in the field of imaginary numbers (i^n), or as a result, in the field of complex numbers, C^n. It is itself a subgroup of the general linear group, SU(n) ⊂ U(n) ⊂ GL(n, R), or similarly, TU(n) ⊂ U(n) ⊂ GL(n, i).

The group SU(2) is isomorphic to the group of quaternions of normalization of 1, and is diffeomorphic to the 3-spatial sphere. A quaternion is a complex number with a three-dimensional imaginary number (temporal) and one real number (spatial) of the form d + ai + bj + ck, where d, a, b, c are real numbers and i, j, k are imaginary units that satisfy certain conditions. The three spatial dimensions are folded and the three temporal dimensions are unfolded in a (1 + 3) spatiotemporal formalism. Consequently, a hyperternion is defined as a complex number with a three-dimensional real number (spatial) and one imaginary number (temporal) of the form di + ax + by + cz, where d, a, b, c are real numbers and x, y, z are real units (spatial) that satisfy certain conditions, where di is an imaginary number (temporal). The three temporal dimensions are folded and the three spatial dimensions are unfolded in a (3 + 1) spatiotemporal formalism. An example of a hyperternion would be the four-dimensional coordinates of a point on a tesseract or hypercube. The group TU(2) is isomorphic to the group of hyperternions of a normalization of 1, and is thus diffeomorphic to the 3-temporal sphere.

A group homomorphism that is surjective reaches every point in the codomain. Since unit quaternions can be used to represent rotations in three-dimensional space, there is a surjective homomorphism from SU(2) to the rotation group SO(3) whose kernel is {+I, −I}, where I is the 3 × 3 spatial identity matrix. SU(2) is also identical to one of the symmetry groups of spinors, Spatial Spin(3), that enables a spinor presentation of rotations. The group SO(3) is used to describe the possible rotational symmetries of an object, as well as the possible orientations of an object during its spin in space. The group TO(3) is used to describe possible rotational symmetries of an object through the temporal volume, or the possible orientation of an object during its tumble in time, whose imaginary conjugate kernel is $\{+I^*, -I^*\}$, where I^* is the 3 × 3 temporal identity matrix.

Hence, for a six-dimensional color string, we denote

$$SU(3) - TU(3) \equiv \pm U(6) \qquad (1.1)$$

For a four-dimensional surface with a (2 + 2) formalism, we have

$$SU(2) - TU(2) \equiv \pm U(4) \qquad (1.2)$$

$$\text{Spatial Spin}(3) - \text{Temporal Spin}(3) \equiv \pm U(4) \qquad (1.3)$$

For a three-dimensional spatial volume, we obtain

$$SU(3) \subset U(3) \subset GL(3, R) \qquad (1.4)$$

For a six-dimensional spatiotemporal volume with a (3 + 3) formalism, we can write

$$[SU(3) \wedge TU(3)] \subset U(6) \subset GL(6, C) \qquad (1.5)$$

Representing SU(6) with a (3 + 3) formalism for six-dimensional space-time in terms of a combination of three spatial dimensions and three pseudo spatial dimensions, we obtain

$$SU(6) \subset U(6) \subset GL(6, C) \qquad (1.6)$$

A general notion of unitary transformations, since unitary groups are just groups consisting of unitary transformations, helps to form a physical intuition behind unitary groups. It is helpful to start with the rotations from classical physics, which are known as orthogonal transformations, that are more familiar objects, to develop a good physical intuition.

An orthogonal transformation is a linear transformation which preserves a symmetric inner product. In particular, an orthogonal transformation preserves lengths of vectors and angles between vectors. Why is that the case? It is because the unitary transformations are basically the same thing as the orthogonal transformations when you start using complex numbers, or complex vector spaces, instead of real numbers.

Hence, rotations in three spatial dimensions are simply transformations that do not change the dot products between vectors.

$$R\vec{q} \cdot R\vec{p} = \vec{q} \cdot \vec{p} \tag{1.7}$$

So, the rotations of classical physics are special because observers with axes that are rotated relative to each other will measure the same physical relations. Moreover, the rotation group is a subgroup of the Lorentz group which connects the measurements of two inertial observers. Likewise, a certain type of transformation may yield a measurement from physical reality that may be utilized to better understand the unitary transformations.

Let us consider a vector space over the complex numbers where there is a notion of "dot product." In such a general context, a product is usually called an inner product, and in that context we can undertake transformations that do not change the inner products between vectors such as:

$$\langle U\vec{q}, U\vec{p} \rangle = \langle \vec{q}, \vec{p} \rangle \tag{1.8}$$

and these types of transformations are called unitary transformations.

Quantum Mechanics has at its foundation complex vector spaces with inner products since the pure states of quantum systems are vectors in complex vector spaces, and the inner products of these vectors allow us to calculate probabilities for certain measurement outcomes. In the above context, unitary transformations become very physical and very significant. They can be regarded as symmetry transformations on the quantum system because they preserve the inner product which determines measurement probabilities. Moreover, unitary transformations do not change measurement outcomes in Quantum Mechanics, in the same way that rotations do not change what an observer measures in classical physics. (Wigner, 1931)

In Quantum Mechanics, the normalization of probabilities is represented by the normalization of state vectors with respect to the wave function that is itself pure probability. Therefore, it is possible to describe the expectation value "E" of an observable in state $|\psi\rangle$ as $\langle\psi|E|\psi\rangle$, and its normalization ensures that the expectation value of a constant is that same constant, $\langle\psi|1|\psi\rangle = \langle\psi|\psi\rangle = 1$. Consequently, this assertion is the requirement that probabilities have to add up to 1.

If a transformation is a time evolution, it would definitely be desirable to ensure that the normalization condition is conserved. Moreover, it would be usually desirable to also have reversible operations such as the transformation involving the time evolution of closed quantum systems, which would be needed for infinite-dimensional Hilbert spaces, but not for finite-dimensional Hilbert spaces where the requirement of conservation of normalization suffices.

Let us now consider the transformation of U with the expectation of a result, for the state that has been transformed through time evolution as $|\varphi\rangle = U|\psi\rangle$. Consequently, the normalization condition becomes $\langle\varphi|\varphi\rangle = \langle\psi|U^\dagger U|\psi\rangle$ for all $|\psi\rangle$, and ensues that, $U^\dagger U = 1$, since U was supposed to be reversible, it indicates that $U^\dagger = U^{-1}$, which confirms the previous definition of a unitary transformation.

Therefore, there are unitary transformations.

§ 2. What is a unitary group?

It is possible to declare that a unitary group is simply a set of unitary transformations with certain properties:

- The unitary group is a subgroup of the general linear group, GL.

- The inverse transformation is encompassed within every transformation. So, as a result, the inverse transformation of a unitary transformation is itself a unitary transformation.

- The unitary group would contain the identity transformation, which could do nothing to change the normalization, which confirms it is a unitary transformation.

- After undertaking two transformations, the result would be the product of the two transformations if two consecutive operations are applied and the end result is again a unitary transformation, with an unchanged normalization.

§ 3. How does a finite unitary group appear in Quantum Mechanics prior to the Quantum Field Theory?

It has been known that the electron has a quantum property called "spin" which is a form of angular momentum, an intrinsic property or quantum property of an electron. When that intrinsic property is measured in any direction, the observer gets one of two values, "spin up" (usually denoted by $|\uparrow\rangle$) or "spin down", which is usually denoted by $|\downarrow\rangle$. Therefore, if the space dependence of the wave function is disregarded, the spin state that is most common for an electron according to the superposition principle is $|\psi\rangle = \mu|\uparrow\rangle + \nu|\downarrow\rangle, |\mu|^2 + |\nu|^2 = 1$, where the condition is on account of the constraint of normalization.

It is interesting to note that the "spin up" state, or the "spin down" state, are defined relative to a certain direction. Nevertheless, nature is isotropic, and nature does not have a preferred direction even though a particular experiment may be designed to establish a specific direction; for instance, by using an external electromagnetic field.

Hence, it is possible to represent the same "spin up" state or the same "spin down" state of any other direction which is equivalent to the rotation of the frame of reference that is involved in this case. For example, let us describe the "spin up" state and the "spin down" state corresponding to those other directions such as the "spin up" $|\uparrow*\rangle$ and the "spin down" $|\downarrow*\rangle$ that can be written as:

$$|\psi\rangle = \mu|\uparrow\rangle + v|\downarrow\rangle = \epsilon|\uparrow*\rangle + \beta|\downarrow*\rangle \qquad (3.1)$$

Clearly, the transformation from (μ,v) to (ε,β) needs to be linear in order to support superpositions. Consequently, the transformation needs to be reversible because the equation can be interpreted bidirectionally. Nonetheless, the normalization conditions have to be complied with in either direction of each interpretation.

Therefore, a reversible linear operation that conserves normalization may be applied to proceed from (μ,v) to (ε,β) that is also a unitary operation. The two-dimensional set of unitary operations is known as U(2), but not all these unitary operations are needed because the physical state is not changed by the global phase on the state vector.

It is interesting to reiterate that unitary transformations always have a determinant with an absolute value of 1, since the phase is insignificant in this context, it may be selected so that the determinant is actually a real value of 1. Hence, the transformations that are known as "special unitary operations" form a group known as SU(2), and all of SU(2) is needed to describe spin rotations.

As previously stated, there is a close relation of the group SU(2) with the group SO(3) of spatial rotations. Thus, for every SU(2) transformation, there is a unique corresponding SO(3) transformation, but not the other way around, because for each SO(3) transformation there are two SU(2) transformations. In fact, those two transformations do not make a physical difference because they only differ in sign. Nevertheless, the minus sign of the transformations cannot be disregarded without profound repercussions, like it may be done with the rest of the phase, when going from U(2) to SU(2).

PART III

THE FRONTIER OF QUANTUM PHYSICS

Chapter 5

The Theory of Color Loop Quantum Gravity (CLQG)

§ 1. Is the Color Spinor and its Color Spin Connection a case for quantum gravity with no color strings attached?

Does Color Loop Quantum Gravity try to quantize General Relativity with no color strings attached?

From previous research, the emergence of a counter-gravitational field has been described in terms of its wave theory, but it could have also been described in terms of spinors. Spinors were introduced in geometry by the eminent mathematician Élie Cartan in 1913. (Cartan, 1981) A spinor is a vector-like quantum of intrinsic angular momentum or spin as connection over a closed loop. A spinor may be conceptualized as the square root of a section of vector bundles; the product of a spinor and its complex conjugate makes a vector. A spinor may be considered a primitive object from where a vector or a tensor can be introduced in a manifold with a metric. For example, during a double-sided transformation of a Pauli vector, $\psi V \psi^{\dagger}$, a spinor only rotates half as much as a vector.

Hence, half a rotation in the state space of a spinor is a full rotation in the physical space of a vector, or a full rotation in the state space equals two rotations in the physical space. Spinor analysis may be considered a substitute for complex analysis. So, a spatiotemporal dimension about a point may connect back on itself, to define any spatial three-dimensional geometry of these closed dimensional loops. Each closed loop is an elementary color graviton with its color spinor that quantizes gravitation, and connects and interlaces the spatiotemporal fabric. Each quantum of gravitation is quantized in a background independent manner to manifest color loop quantum gravitation. Thus, a color spinor would rely on the emergence of a color graviton.

To keep the subject elementary, the author has stated without proof the theorem of Cartan on linear representations of simple groups from previous research. (Nieves, 2020) They provide a linear representation of the group of rotations in a space with any number "n" of dimensions; each spinor having 2^v components where $n = 2v + 1$ or $2v$. Spinors in six spatiotemporal dimensions occur in the six-dimensional electrogravitic Dirac equations, the six wavefunctions being nothing other than the components of a spinor.

Let us consider the spatiotemporal medium of special relativity referred to coordinates $x^1, x^2, x^3, x^4, x^5, x^6$, a function of position f. The differential df is an invariant scalar under all direct or inverse Lorentz transformations.

The covariant vector $\partial/\partial x$ may be expressed as

$$df \equiv \frac{\partial f}{\partial x^1}dx^1 + \frac{\partial f}{\partial x^2}dx^2 + \frac{\partial f}{\partial x^3}dx^3 + \frac{\partial f}{\partial x^4}dx^4 + \frac{\partial f}{\partial x^5}dx^5 + \frac{\partial f}{\partial x^6}dx^6 \quad (1.1)$$

The differential dx^i transform as the components of a contravariant vector and consequently we can regard the six operators $\partial/\partial x^i$ as the components of a covariant vector; the contravariant components of this vector are given by

$$\frac{d}{dx^1}, \frac{d}{dx^2}, \frac{d}{dx^3}, -\frac{1}{c}\frac{d}{dx^4}, -\frac{1}{c}\frac{d}{dx^5}, -\frac{1}{c}\frac{d}{dx^6} \quad (1.2)$$

A spinor describes rotation at a specific spatiotemporal point independently of rotation at any other point in space-time.

$$\vec{\Psi}(r,t) = \sum_{n=1}^{6} \vec{\Psi}_n(r,t) = \begin{vmatrix} \vec{\Psi}_1(t) \\ \vec{\Psi}_2(r) \\ \vec{\Psi}_3(t) \\ \vec{\Psi}_4(r) \\ \vec{\Psi}_5(t) \\ \vec{\Psi}_6(r) \end{vmatrix} \quad (1.3)$$

Denoting the associated matrix as $\vec{\mathfrak{R}}$ instead of $\partial/\partial x$ from previous research, where $\vec{\mathfrak{R}}$ is the Robertonian six-dimensional operator, and $\vec{\Psi}(r,t)$ is the six-dimensional spatiotemporal wavefunction spinor of the system.

$$\vec{\mathfrak{R}} = -\frac{1}{c}\frac{\partial}{\partial t_x}\vec{a}_{t_x} + \frac{\partial}{\partial x}\vec{a}_x - \frac{1}{c}\frac{\partial}{\partial t_y}\vec{a}_{t_y} + \frac{\partial}{\partial y}\vec{a}_y - \frac{1}{c}\frac{\partial}{\partial t_z}\vec{a}_{t_z} + \frac{\partial}{\partial z}\vec{a}_z \quad (1.4)$$

Let us introduce the six wavefunctions that are the components of a spinor $\vec{\Psi}(r,t)$ and functions of position "x"; and V be the associated vector potential.

$$V = c\left(\sum_{n=1}^{3} \alpha_n \hat{p}_n\right) \quad (1.5)$$

The temporal six-component Dirac-Lorentz matrix β is given by

$$\beta = \begin{vmatrix} 1 & 0 & 0 & 0 & 0 & 0 \\ 0 & 1 & 0 & 0 & 0 & 0 \\ 0 & 0 & -1 & 0 & 0 & 0 \\ 0 & 0 & 0 & -1 & 0 & 0 \\ 0 & 0 & 0 & 0 & 1 & 0 \\ 0 & 0 & 0 & 0 & 0 & 1 \end{vmatrix} \quad (1.6)$$

We may express the energy difference of the above two left terms as a single energy expression, with six-component Dirac-Pauli temporal-spin matrices, or gamma matrices, γ^ε, and a three-dimensional spatial momentum matrix, $\hat{p}_n$, with six components. (Nieves, 2020)

With this notation the six-dimensional Dirac equations for an electron in an electromagnetic field are as follows:

$$\left\{ i\hbar\vec{\mathfrak{R}} - c\left(\sum_{n=1}^{3}\alpha_n \hat{p}_n\right) - \beta m'c^2 \right\}\vec{\Psi}(r,t) = 0 \qquad (1.7)$$

Simplifying the relativistic equation, we have,

$$i\hbar c \gamma^\varepsilon \vec{\mathfrak{R}} = i\hbar\vec{\mathfrak{R}} - c\left(\sum_{n=1}^{3}\alpha_n \hat{p}_n\right) \qquad (1.8)$$

$$\left(i\hbar c\gamma^\varepsilon \vec{\mathfrak{R}} - \beta m'c^2\right)\vec{\Psi}(r,t) = 0 \qquad (1.9)$$

Where the symbols: *i, ℏ, c, and m′*, all have well known physical meanings.

The relativistic six-dimensional quantum mechanical wave equation, including electromagnetic interactions, describes all spin−½ massive particles for fermions (all quarks and leptons), that are symmetric under parity, or symmetric if the sign of one spatial coordinate is flipped. This equation is consistent with the Special Theory of Relativity and the Principles of Quantum Mechanics and includes the evolution of three-dimensional time. The equation encompasses six wave equations of motion for an electron, a positron, an electron neutrino, and their anti-particles, submerged in an external electromagnetic field in six-dimensional space-time.

The relativistic six-dimensional equation has six components or states, or six degrees of freedom, for particles and antiparticles, each component is a direction of spin or anti-spin. As predicted by Dirac, each particle, or antiparticle, is always moving at *"c"* with a trembling motion, $\langle v \rangle = \pm c$, due to weaker Coulombic forces near protons at Compton wavelength distances, which makes the motion appear slower, even though the motion abides by the Special Theory of Relativity. The single relativistic six-dimensional equation unfolds into six coupled linear first-order partial differential equations for the six components that make up the six-dimensional quantum mechanical wave function.

Aside, per Cartan's observation, in a spatiotemporal medium with an odd number of dimensions, systems of equations to Dirac's equations that are invariant under reversals and displacements do not exist; this follows from the fact that the spinor of the former is not equivalent to the spinor of the latter with respect to reversals. Nonetheless, in a spatiotemporal medium with whatever even number of dimensions, Dirac's equations generalize as they stand. (Cartan, 1981).

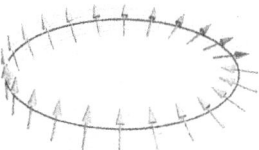

Figure 1. An Illustration of A Spinor.

Therefore, the spatiotemporal landscape, or the background of elementary particles, may be conceptualized as a porous medium of color pixels, or spatiotemporal point sources, that cover the entire fabric of space-time of physical reality, like an elementary closed circuit of gravitational field in a background independent way.

This realization presents an alternative representation of the General Theory of Relativity in terms of color spinors and color gravitons, quanta of gravitation, as its constituents. The color gravitational loops may be interlaced into a spin-network using Ashtekar variables for a large-scale smooth spatiotemporal landscape. However, at a quantum scale, the spatiotemporal landscape would be wavy, quantized with spatiotemporal facets, and pixelated due to the spatiotemporal emergence or convergence at any arbitrary point. (Ashtekar, 1986)

It is possible to theorize that Color String Theory, Quantum Mechanics, and General Relativity, may be combined without supplanting or removing their key foundational principles. Color Loop Quantum Gravity has background independence in six-dimensional space-time or in folded four-dimensional space-time since color strings are spatiotemporal in nature. Space endows time, and then time endows more space. Time is addressed by "A Dynamic Theory of Space-Time". (Nieves, 2020) The combined theoretical framework would predict current predictions of each of its components, including the six-dimensional Einstein field equations, and the quantum mechanical equations. The speed of light depends on the spatiotemporal waves due to the expansion or contraction of space-time at every point on its path. Color Loop Quantum Gravity emerges from Color String Theory.

All there is has a spatiotemporal aspect, and space-time may expand, stay static, or contract, as it has been conceptualized, so, color strings or any other physical manifestation has the same spatiotemporal properties. The red-shift or blue-shift of light as it travels through the spatiotemporal medium is due to the wave property of the spatiotemporal medium. The Higgs boson has been detected, not observed, as any other potential quantum manifestation may be eventually detected if it exists in our physical reality or virtuality.

Consequently, color strings vibrate within a certain spatiotemporal medium as a form of energy, but color gravitons may endow gravitational fields that endow Higgs fields. So, the Color String Theory underpins the Gluon Standard Model of Quantum Physics.

General Relativity describes gravity in terms of interactions between mass and space-time. Space-time is dynamic and background independent. Quantum Mechanics describes the composition and behavior of matter in terms of elementary constituents. Space-time is passive and background dependent. Color String Theory describes all of the above and can be both spatial background dependent and temporal background independent.

§ 2. Could space-time be the source of both a dependent and an independent background of physical reality as well as the quintessential source of all there is?

Let us imagine the color graviton, or color string loop, as being created around a color pixel or spatiotemporal source point. The Color Graviton forms around the pixel point where it may stay or propagate depending on its frequency, energy, or interaction with other color charges or color gravitons, or the perturbations in its spatiotemporal background.

For example, a torus of ice may form around a water source point by some method where the water's temperature is cold enough not to cause the donut-shaped body of ice to melt, but not too cold to freeze the water around it. Hence, the torus of ice and the water are made of the same substance but they do not have the same phase of matter. The torus is solid and the water around is a liquid. It is possible to conceptualize that the water around the torus is its separate background or medium, and the torus may be independent of its background as a persisting system of mass. But it is also the case that both the object and its background consist of the same substance but in a different physical phase of existence and with other different properties.

The color string loop, or color graviton, may have an analogous physical manifestation in our reality. The spatiotemporal background of the color graviton may consist of color pixels or spatiotemporal points, and there may also be spatiotemporal source points in the spatiotemporal background, where space endows time, and then time endows more space. The Color Graviton as a color string loop may propagate through six-dimensional space-time or it may also travel forward or backward through time on its retarded wavefunction or on its advanced wavefunction in the physical dimensions of reality. The wavefunction and its position and momentum operators are tied to the spatial coordinate system to describe the location and change of location over time, which makes the wavefunction highly background dependent. But the wavefunction temporal operators, as given in A Dynamic Theory of Space-Time, are not tied to the spatial coordinate system, so the temporal wavefunction is highly background independent.

The Color Graviton travels on the retarded or advanced temporal wavefunction at the speed of light. Thus, there may be different spatiotemporal manifestations that may exist in the independent background of space-time as previously stated. A theory of color loop quantum gravity solves the conundrum of the dependence versus independence of the spatiotemporal background for elementary particles or systems of particles. It describes the quantum evolution of the six-dimensional geometry of space-time.

A Color Graviton traveling on the retarded wave is a type of luxon, as is a gluon, a photon, and any particle with zero rest mass. Tachyons are particles with imaginary rest mass. A color graviton traveling on the advanced wavefunction is a type of tachyon. In color string theory, a theory of color loop quantum gravity, the color graviton is a massless state of an elementary color string. The Color Graviton is predicted to be massless because its gravitational force is very long range and it propagates at the speed of light.

A color graviton is both a virtual particle and a gravitational wave. A color graviton may be described as a contracted spatiotemporal volume while a color anti-graviton may be described as an expanded spatiotemporal volume with opposite rotation or torsion within its particle boundary. A gravitational wave may encompass color gravitons within its wavelength and frequency. A color graviton is a nubble of gravitation that manifests the quantum gravitational state of a virtual relativistic mass.

§ 3. Are the Higgs Scalar Boson and the Higgs Field part of the Color String Theory?

The Higgs boson is an elementary particle produced by the quantum excitation of the Higgs field according to the Gluon Standard Model. The Higgs scalar boson is a particle of mass with no electrical charge, with zero spin, and no color charge. The Higgs scalar boson is very unstable, and it may decay into other quantum particles soon afterward. Thus, it is quantum in nature as it breaks down into its constituents that consist of elementary particles themselves. So, the Higgs particle has a fundamental substructure. Is it possible that the Higgs scalar boson has a color string substructure? Could the energy density of the color strings and color gravitons be the source of the Higgs particle? Could elementary particles exchange color gravitons similar to the way they exchange gluons? It is possible to theorize the substructure for the Higgs scalar boson based on the properties of the color graviton, color strings, and gluons.

It is possible to think about the Higgs mechanism as a field of the common burdock plants with stickseeds or hitchhikers, also known as burs; a seed or dry fruit that has hooks or teeth, that stick to your clothes or to an animal's fur very easily, as a person or an animal walks through the field. Moreover, the more a person or an animal walks through the field the more stickseeds hitch a ride and the greater the mass that a person or animal gains. The Higgs mechanism was proposed by the eminent physicist Peter Higgs in 1964 along with renowned scientists Robert Brout, and François Englert. The Brout-Englert-Higgs mechanism explained why some quantum particles acquire mass, but it does not explain yet how and why all particles have mass. At the time, there was no evidence of a Higgs particle or field. A subatomic particle with the characteristics of the Higgs particle was discovered in 2012 at the Large Hadron Collider at CERN.

The color mass acquisition is crucial to the generation of the property of mass for all particles and systems of mass. However, the Higgs field is colorless, that is it does not have a net color charge, but the field could very well have color charges that offset one another. From the gluon constituents of the color field emerge the masses of the fundamental quarks, leptons and gauge bosons. Let us describe the color charge contraption for this fundamental mass acquisition process that encompasses the Higgs mechanism. The Color Charge Contraption is the process that generates the mass of approximately 99% of composite particles like baryons, such as nucleons, from quantum fields of the gluons, the binding energy of gluons, according to the Gluon Standard Model as previously explained in "A Dynamic Theory of Space-Time". (Nieves, 2020) This gluonic energy is the sum of the energies of the massless gluons mediating the strong interaction inside the baryons and of the kinetic energies of quarks.

The color charge contraption also endows the property of mass to quantum particles as a transfer of potential energy to the particles from the Higgs field as the particles couple with the field, or in layman's terms, as the Higgs particles hitch a ride on the quantum particles. The Higgs field is proposed to contain the mass property in the form of colorness or color energy from gluons, quarks, and their fields. It is also important to mention that the relativistic speed of a massive or massless particle would also add relativistic mass to the rest mass of a particle due to its motion and the properties of the spatiotemporal medium.

The color energy of the Higgs field may also consist of color gravitons that bind with anticolor gravitons to make long chains of color charges, or color charge polymers, through entropic forces, which build more massive colorless structures that are attracted to, or gravitate toward, the color charges of the hadrons, but do not bind with them because of their colorlessness, or overall weak color charge of the long color graviton structures, and the topological properties of the color string loops.

A Hadron is a quantum particle that includes the mesons and baryons, which takes part in the strong interaction between color strings or gluons. A Hadron is a composite particle made of two or more quarks bound together by the strong color force. Thus, the Higgs field is crucial in the generation of the masses of quarks, leptons, $W^\pm$ and Z^0 gauge bosons, through the Yukawa coupling mechanism. The gauge bosons and the scalar boson have sufficient energy to decay; so, any spin-zero particle decays into a lighter particle. Even the theoretical color graviton is predicted to decay to two spin-1/2 or two spin-1 particles.

3.1 The Relationship between Geometry and Energy Momentum for the Spatiotemporal Medium of Color Gravitons.

Could color gravitonic polymers change their geometry due to a change in entropy after a heat transfer?

The Raychaudhuri equation demonstrates how gravitation can be attractive in the General Theory of Relativity. This demonstration is done by considering a bundle of geodesics, and proving that the time derivative of the spatiotemporal expansion of an area "θ" is negative under certain reasonable conditions. Hence, the bundle of geodesics tends to converge with the passage of time while the spatiotemporal curvature manifests an attractive gravitational force. (Raychaudhuri, 1955)

In general, the Raychaudhuri equations are concerned with the kinematics of flows. In particular, the values of the evolution equations indicate the characteristics along the flow through a specific spatiotemporal background. One of these evolution equations is known as the expansion equation or the Raychaudhuri equation. The parameter "λ" indicates points along the curves of a flow, to characterize the flow with distinct functions of "λ". The gradient of the velocity field of the flow has three terms: the traceless part, the symmetric part, and the anti-symmetric part. These parts of the flow are the expansion, shear, and the rotation. Below is an illustration of an area enclosing a set of lines of flow.

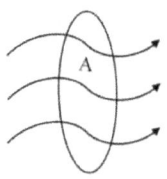

Figure 2. The cross-sectional area enclosing a bundle or congruence of geodesics.

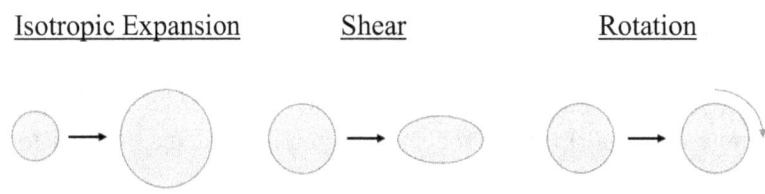

Figure 3. An Illustration of the expansion, shear, and rotation of a spatiotemporal area.

$$\text{Heat Transfer} \equiv \text{Transport of Energy-Momentum} \quad (3.1)$$

$$\delta Q = -\kappa \int \lambda T_{\alpha\beta} k^{\alpha} k^{\beta} d\lambda \delta A \quad (3.2)$$

$$\delta A = \int \theta d\lambda dA \quad (3.3)$$

In General Relativity, the Raychaudhuri equation describes the fundamental motion of particles or polymers of mass that are near. The equation provides a simple as well as a general proof that gravitation should be attractive in our universe between any two systems or polymers of ordinary mass-energy according to the General Theory of Relativity, and for the analysis of exact solutions in General Relativity. Moreover, the equation is an essential intermediate theorem for the Penrose-Hawking singularity theorems.

From previous research, a spatiotemporal point is zero-dimensional, or mathematically equal to the magnitude of the square root of an ordinary vector. The quantum effect of a color graviton may be related to this infinitesimal-scale structure of the spatiotemporal medium and its quantum gravitational effects. Space-time is significantly curved around a spatiotemporal point source to manifest a color string loop. So, the spatiotemporal expansion rate and the spatiotemporal metric may be a very distinctive medium for bundles of null geodesics that are time-like.

The flow of bundles of null geodesics through a spatiotemporal manifold is described by the expansion equation that may explain the formation and confinement of color string loops, color charge combinations and spatiotemporal singularities. The expansion equation is purely geometrical in nature to explicate kinematic entropy variation. It is the component of the Ricci tensor of the expansion equation that provides a connection to the General Theory of Relativity as a classical gravitational theory.

The gradient of the Raychaudhuri equations characterizes the velocity field. The gradient of the velocity field indicates the path where the field changes the most. If the EFEs are applied as a geometric background, these equations become a coupled system of equations to the characteristics of the spatiotemporal expansion, shear, and rotation, with the specified initial conditions, for the null geodesic flow. Consequently, it may be possible to find the velocity field and its gradient.

From the field of optics, it has been shown that a caustic, or a caustic network, is the envelope of light rays refracted or reflected by a curved surface or an object, or the projection of that envelope of light rays on another surface. The caustic is a curve or surface to which each of the light rays is tangent, defining a boundary of an envelope of light rays as a curve of concentrated light. For example, the lit up edges on the surface of ocean waves are made by caustics.

As the EFEs are applied with acceptable assumptions on the stress–energy– momentum tensor, the expansion equation may describe how the null geodesic congruences converge to form caustics. This potential result is due to the fact that as the null geodesic congruences converge they lead to attractive gravitation. It is interesting to point out that under certain conditions these caustics may form cosmological singularities or black holes. A theory of quantum gravity like "A Dynamic Theory of Space-Time" would manage these potential spatiotemporal singularities.

To find the convergence or divergence of null geodesics, we have,

$$\frac{d\theta}{d\lambda} = -\left(\frac{1}{n-1}\right)\theta^2 - \sigma^2 + \omega^2 - R_{\alpha\beta}k^\alpha k^\beta \qquad (3.4)$$

Thus, rotation "ω^2" defies convergence, while shear assists it. The equation for the evolution of the rotation $\omega_{\alpha\beta}$, has a trivial solution given by $\omega_{\alpha\beta} = 0$. The criterion for convergence then becomes particularly simple for such hypersurface orthogonal congruences (with zero rotation): $R_{\alpha\beta}k^\alpha k^\beta \geq 0$. This leads to what is known as geodesic focusing which conveys the concept that if matter is attractive, the null geodesics must eventually converge. This concept is proved via the focusing theorem.

$$\therefore \delta A = -\int \lambda R_{\alpha\beta}k^\alpha k^\beta d\lambda \delta A \qquad (3.5)$$

Where "σ^2" is a geodesic distance, k^α and k^β are null geodesic congruences (bundles of null geodesic curves) that are time-like, of two hyperplanes "α" and "β", both orthogonal to a time-like unit vector $\vec{k}$, "θ" is the trace of the expansion tensor $\theta_{\alpha\beta}$ of null geodesics, $R_{\alpha\beta}$ is the Ricci tensor, the term $\left(\pm R_{\alpha\beta}k^\alpha k^\beta\right)$ is a quantity sometimes called the Raychaudhuri scalar, or the trace of the tidal tensor (the plus sign is for time-like curves whereas the negative sign is for space-like curves), "λ" denotes the parameter labeling points on the curves in the flow, "n" is the number of spatiotemporal dimensions, and "A" is the geometrical area of expansion, shear, or rotation if any.

Aside, since rotation defies convergence, it is possible to ask the rhetorical questions: if the spatiotemporal area on the event horizon of a supermassive black hole rotates at the speed of light, could rotation ever exceed expansion and shear on bundles of null geodesics of the event horizon? Could a universe rotate in a way that could exceed expansion and shear? Would such a universe converge or diverge? Could the eminent mathematician Kurt Gödel have been always right? when he repeatedly asked, "Is the universe rotating yet?"

What is the relationship between entropy variation, mass-energy density and spatiotemporal curvature variation? Is it similar to the relationship between the mass-energy density of the Einstein Field Equations and the spatiotemporal geometry?

$$R_{\alpha\beta}k^\alpha k^\beta = -\left(\frac{1}{n-1}\right)\theta^2 - \sigma^2 + \omega^2 - \frac{d\theta}{d\lambda} \quad (3.6)$$

Applying the null geodesic congruences $k^\alpha k^\beta$ to the EFEs,

$$R_{\alpha\beta} - \left(\frac{1}{n-1}\right)Rg_{\alpha\beta} + \Lambda g_{\alpha\beta} = \frac{8\pi G}{c^4}T_{\alpha\beta} \quad (3.7)$$

$$R_{\alpha\beta}k^\alpha k^\beta - \left(\frac{1}{n-1}\right)Rg_{\alpha\beta}k^\alpha k^\beta + \Lambda g_{\alpha\beta}k^\alpha k^\beta = \frac{8\pi G}{c^4}T_{\alpha\beta}k^\alpha k^\beta \quad (3.8)$$

$$R_{\alpha\beta}k^\alpha k^\beta = \frac{8\pi G}{c^4}T_{\alpha\beta}k^\alpha k^\beta + \left(\frac{1}{n-1}\right)Rg_{\alpha\beta}k^\alpha k^\beta - \Lambda g_{\alpha\beta}k^\alpha k^\beta \quad (3.9)$$

For a system or a polymer of ordinary mass-energy that is tangential to bundles of null geodesics through the flow, we have,

Kinematic Entropy Variation $\equiv \pm$ (Mass-Energy Density + Spatiotemporal Curvature) (3.10)

$$-\left(\frac{1}{n-1}\right)\theta^2 - \sigma^2 + \omega^2 - \frac{d\theta}{d\lambda} = \frac{8\pi G}{c^4}T_{\alpha\beta}k^\alpha k^\beta + \left(\frac{1}{n-1}\right)Rg_{\alpha\beta}k^\alpha k^\beta - \Lambda g_{\alpha\beta}k^\alpha k^\beta \quad (3.11)$$

The kinematic entropy describes the characteristic quantities in the flow of a system of mass-energy particles along the bundles of null geodesics that may converge or diverge, given by the variation in the sum of the stress–energy–momentum tensor, the spatiotemporal curvature, and the cosmological constant term. The above variation equation begs the rhetorical question, would this contribution to kinematic entropy also offset the missing fermionic mass?

What is the n-dimensional multiplier of the Einstein gravitational constant?

Let us describe the fraction $\left(\dfrac{1}{n-1}\right)$ for an *n*-dimensional manifold in the Einstein gravitational constant of the *Einstein* field equations, where the "*n*" represents the number of spatiotemporal dimensions. In a six-dimensional spatiotemporal medium, "*n*" would be equal to 6. In the Einstein field equations the n-dimensional fraction multiplies the Einstein gravitational constant "κ" which is equal to $8\pi G/c^4$. If $n = 6$. the *n*-dimensional fraction is equal to ⅕.

A spatiotemporal point in the medium *P(x, y, z)* has two opposite spatiotemporal waves along every spatiotemporal dimension, one of the waves is the retarded wave and the other is the advanced wave. For a six-dimensional spatiotemporal point there are (3) spatial dimensions and (3) temporal dimensions in a (3 + 3) formalism. Each dimension has two directions. The retarded wave travels in one direction and the advanced wave travels in the other direction.

Every axial direction has spatiotemporal components, a spatial component from a spatial axis and a parallel temporal component at quadrature with the spatial axis of another orthogonal dimension. Hence, every spatial dimension has an orthogonal conjugate temporal dimension in space-time. Each dimension has both spatial and temporal dimensions along its directions. Nonetheless, these dimensions are extended along those directions, they are not folded as in the case of the original EFEs. Time is nonlinear. Consequently, $n = 6$ with a (3 + 3) formalism, but not $n = 4$, as in the case of folded temporal dimensions for linear time with a (3 + 1) formalism as previously explained. (Nieves, 2020)

Let us consider a spatiotemporal point in a gravitational field very close to a celestial body, like a planet. One of the dimensions passing through the point is in the radial direction toward the center of the planet. Let us call that the inward direction along that dimension the inward gravitational field direction, or the direction of the –y axis in a Cartesian coordinate system, that yields the negative gravitational field acceleration at the planet. All the other dimensional directions are away from the planet, or opposite to any potential cosmological counter gravitational field. From previous research, it was hypothesized that there would be a cosmological gravitational acceleration that is represented by the Weyl cosmological tensor,

C_{abcd}, and its associated cosmological curvature, counteracting the local curvature of space-time represented by the Riemann curvature tensor, R_{abcd}, the local semi-traceless part, E_{abcd}, (containing the Ricci curvature tensor), and the local scalar part, S_{abcd}.

Therefore, all the components of the positive gravitational forces at point P that are facing outward are acting on the n-dimensional manifold produced by the local spatiotemporal curvature minus the cosmological curvature above the point, except the composite and inward negative gravitational force below the point and toward the center of the planet. If the components of all directional forces in the top hemisphere above the point are projected on the five axes: +x, +y, +z, −x, −z, the resultant force would act against the cosmological curvature. These five axial directions contribute a composite counter gravitational force, not an anti-gravitational force, due to the spatiotemporal expansion at point P. If the counter gravitational force acting on the n-dimensional manifold is equally divided among the five axial directions, the result would be an n-fractional magnitude of force per energy density. (Nieves, 2021)

$$\text{Pressure} \equiv \text{Energy Density} \qquad (3.12)$$

If $n = 4$, then, the Einstein field equations will still deliver a correct amount of energy density for the case of folded six-dimensional time, but the pressure amount would be skewed, which may lead to missing mass or speculation for other unknown forms of energy. However, if $n = 6$, the n-dimensional fraction of ⅕ is utilized for only one fifth of the spatiotemporal curvature for the amount of the energy density in the six-dimensional EFEs. It is interesting to note that the formalism of the source of energy in the six-dimensional stress-energy-momentum tensor is $(3\rho + 3p)$. This realization may lead to a better assessment for the existing missing mass in our universe.

3.2 The Hidden Thermodynamics of the Graviton and its Gravitational Wave

The gravitational wave theory is a hidden-variable theory that has realism, for its concepts exist independently of the observer, and it also has determinism. The positions of the gravitons are considered to be the hidden variables. (de Broglie, 1927)

An action, in physics, describes how a physical system changes over time. Consequently, the equations of motion can be derived through the principles of least action. For a single graviton moving with a specific velocity, the action of the graviton is its momentum times the total distance traveled along its trajectory, or twice its kinetic energy times the temporal period for which the graviton has the same amount of energy. For complex systems, all such quantities are summed up.

Entropy is the thermodynamic property of a substance at phononic equilibrium, or a measure of the number of possible phononic states of a thermodynamic system in thermal equilibrium. Entropy is a consequence of the passage of time and the expansion of space-time. Entropy (S) is a measure of the equilibrium of energy in a thermodynamic system. Entropy is a measure of unexpected changes that tend to average out, or smooth out, differences in temperature, pressure, density, and chemical potential that may exist in a thermodynamic system. Entropy and Enthalpy are proportional to each other. A change in Enthalpy (H) is proportional to a change in Entropy per unit of absolute temperature.

Entropy is a derived construct of Enthalpy. Therefore, Enthalpy is really the true physical attribute of the system; that is, a change in the energy state of the system. Enthalpy is intrinsic energy. The work that the thermodynamic system is capable of doing is directly related to the enthalpy of the system. Hence, the change of Entropy of the thermodynamic system is equivalent to the change of Enthalpy of the system per unit of absolute temperature, $\partial S = \partial H / T$. As a consequence, the gravitational wave theory brings back the uncertainty principle to distances around extrema of action, distances corresponding to reductions in Entropy or Enthalpy.

The hidden thermodynamics of isolated gravitons are based upon the principles of least action through space or time and Carnot's principles for the heat transfer efficiencies of a thermodynamic cycle.

The action of a graviton becomes the inverse of Entropy or Enthalpy, through an equation like:

$$\frac{Action}{h} = -\frac{Entropy}{k_B} \equiv -\frac{Enthalpy}{m_k v_k^2} \qquad (3.13)$$

$$Action \cdot m_k v_k^2 \equiv -Enthalpy \cdot h \qquad (3.14)$$

where "k_B" is the Boltzmann constant, equal to 1.38065×10^{-23} J/K, "h" is Planck's constant, equal to $6.62607004 \times 10^{-34}$ (Kg·m²)/s, and Energy is represented by the energy of a graviton, $m_k v_k^2$.

In the de Broglie–Bohm concept of quantum mechanics, empty waves can exist as wave functions propagating in space-time, but not carrying momentum or energy, and not associated with a particle or a system of particles. (Selleri et Alia, 1990) (Bohm, 1952)

3.3 The Klein-Gordon Equation for an Effective Field Theory of the Color Graviton.

The Klein-Gordon equation may be regarded one of the most fundamental equations of quantum gravity field theory, and an equation for the motion of a pion, or a free pseudoscalar field or infinitesimal dispersion from a color graviton, of possibly non-vanishing mass "m" on a Lorentzian manifold in a possibly curved spatiotemporal medium. The linearization of the EFEs indicates that small perturbations of the metric, or dispersive gravitational waves, obey a Klein-Gordon-type equation, as the waves propagate at the speed of light.

A color graviton is expected to have an infinitesimal mass, $\sim 1.944 \times 10^{-60}$ Kg. Consequently, the motions of color gravitons can generate kinetic energy. Hence, they have both energy and mass, obey the law of conservation of energy and matter, and explain the phenomenon of the gravitational force at a distance to some extent. Other particles have mass and are much larger than the color graviton, but much less numerous, and they do not manifest the gravitational effects that generate space curvature. If the color graviton has mass, its quantum interaction is at a quantum distance from mass, but classical gravitation obeys the inverse-square at classical distances, without decreasing faster with distance, depending on the masses of the color gravitons that mediate the gravitational force near a mass, if the system is treated as a classical system of mass. Color Gravitons are not the only gravitational mechanism since gravitation may also be generated by the spatiotemporal geometry around a system of mass. The color graviton can decay to two spin-1/2 or two spin-1 particles.

Hence, the detection of a color graviton remains crucial to the validation of a quantum gravity theory and its research that attempts to couple color string theory and quantum mechanics. The extremely weak character of the gravitational interaction makes the detection of a color graviton an extremely difficult task to demonstrate that the gravitational boson mediates the gravitational force. Nonetheless, there is greater hope and probability for polymers of color gravitons to be detected as technology advances.

Using the Klein-Gordon equation, we may obtain the mass of the color graviton, in terms of the infinitesimal color string length "ℓ_s", that can be defined as

$$m_{\text{color graviton}} \equiv \frac{\hbar}{c}\sqrt{\frac{2}{3}\left(\frac{1}{\ell_s}\right)^2} \qquad (3.15)$$

The Klein-Gordon equation is a relativistic equation with inhomogeneity for an effective field theory of gravitation. Effective field theories have found use in General Relativity as they simplify calculations, to allow treatment of gravitational radiation effects, in particular in calculating the gravitational wave signature of inward spiraling finite-sized objects.

An effective field theory is a type of approximation, for a fundamental theory of physical reality, such as the color graviton theory as a quantum field theory model. An effective color graviton field theory includes the relevant degrees of freedom to describe the gravitational phenomenon at the quantum scale, while disregarding substructure and degrees of freedom at infinitesimal distances or at higher energies.

The six-dimensional Klein–Gordon equation is a linear homogeneous second-order partial differential equation with constant coefficients as follows:

The four-dimensional Klein-Gordon equation,

$$\left(\frac{\partial^2}{\partial x^2}+\frac{\partial^2}{\partial y^2}+\frac{\partial^2}{\partial z^2}-\frac{1}{c^2}\frac{\partial^2}{\partial t^2}-\mu^2\right)\phi=0 \qquad (3.16)$$

The six-dimensional Klein-Gordon equation,

$$\left(\frac{\partial^2}{\partial x^2}+\frac{\partial^2}{\partial y^2}+\frac{\partial^2}{\partial z^2}-\frac{1}{c^2}\frac{\partial^2}{\partial t_x^2}-\frac{1}{c^2}\frac{\partial^2}{\partial t_y^2}-\frac{1}{c^2}\frac{\partial^2}{\partial t_z^2}-\mu^2\right)\phi=0 \qquad (3.17)$$

where $\phi(x, t)$ is a pseudo-scalar function, that may be complex in the general case, "m_0" is the rest mass of the particle, and "μ^2" is the spatiotemporal curvature given by $\mu = m_0 c/h$, the reciprocal of a spatial distance.

The Klein–Gordon equation may describe neutral pseudo-scalar particles when "ϕ" is a real function or it may describe charged pseudo-scalar particles when "ϕ^*" is a complex function.

If "ϕ^*" is a complex function, we have,

$$\left(\frac{\partial^2}{\partial x^2}+\frac{\partial^2}{\partial y^2}+\frac{\partial^2}{\partial z^2}-\frac{1}{c^2}\frac{\partial^2}{\partial t_x^2}-\frac{1}{c^2}\frac{\partial^2}{\partial t_y^2}-\frac{1}{c^2}\frac{\partial^2}{\partial t_z^2}-\mu^2\right)\phi^*=0 \qquad (3.18)$$

The interaction of a pseudo-scalar particle, like a color graviton, with an electromagnetic field may be described by the minimal replacement of $\partial/\partial x^i$ for $(\partial/\partial x^i) - ieA_i/\hbar$.

$$\left(\left[\frac{\partial^2}{\partial x^2}-\frac{ieA_x}{\hbar}\right]+\left[\frac{\partial^2}{\partial y^2}-\frac{ieA_y}{\hbar}\right]+\left[\frac{\partial^2}{\partial z^2}-\frac{ieA_z}{\hbar}\right]-\frac{1}{c^2}\left[\frac{\partial^2}{\partial t_x^2}+\frac{\partial^2}{\partial t_y^2}+\frac{\partial^2}{\partial t_z^2}\right]-\mu^2\right)\phi=0 \qquad (3.19)$$

For particles of any spin, including a spin-2 boson, each component of the particle's wavefunction satisfies the Klein–Gordon equation, but only for the case of the Higgs particle, with a spin of "0", the function is invariant with respect to the Lorentz–Poincaré group. According to this group, the four-momentum of a given particle would be invariant, and this was how the concept of internal space-

time symmetries of relativistic particles was formulated.
The Klein–Gordon equation can also be derived by replacing variables for the momentum "*p*" of the particle and for the energy "*E*" from the Special Theory of Relativity, by operators,

$$\frac{1}{c^2} E^2 - p_x^2 - p_y^2 - p_z^2 = m_0^2 c^2 \qquad (3.20)$$

$$E \to -\frac{\hbar}{i} \frac{\partial}{\partial t} \qquad (3.21)$$

$$p_r \to \frac{\hbar}{i} \frac{\partial}{\partial r} \qquad (3.22)$$

The fundamental solutions or propagators of the Klein Gordon equation are found throughout quantum field theories for relativistic perturbations. The equations of motion of a field vector bundle contain the structure of the Klein-Gordon equation, as they do for simpler scalar fields.

The Klein-Gordon equation is the differential equation on smooth functions such as $\phi: X \to \mathbb{R}$ on a pseudo-Riemannian manifold of a spatiotemporal medium (X, g) with a real number $m \in \mathbb{R}_{\geq 0}$, that can be written as

$$\left(\Box_g - \left(\frac{m_0 c}{\hbar} \right)^2 \right) \phi = 0 \qquad (3.23)$$

This equation of motion of the free scalar field on X, of mass "m_0", is in a gravitational field background as incorporated in the metric "*g*", where "$m_0 c/\hbar$" is for the purpose of pure partial differential equation theory just a real number for a spatial distance, the wave operator "$\Box_g$" on the spatiotemporal medium (X, g) is the analog of the Laplace operator in Lorentzian geometry, and "m_0" is the mass of the inverse Compton wavelength. The Minkowski space-time of a spatiotemporal medium like $(X, g) = \mathbb{R}^{p,1}$ is prepared with its canonical functions $x^0 = ct$ and $\{x^i\}_{i=1}^p$, exemplifies the Klein-

Gordon equation and the six-dimensional metric tensor of six-dimensional Minkowski space, "$\eta^{\mu\nu}$" given by

$$\left(\eta^{\mu\nu} \frac{\partial}{\partial x^\mu} \frac{\partial}{\partial x^\nu} - \left(\frac{m_0 c}{\hbar} \right)^2 \right) \phi = 0 \qquad (3.24)$$

Hence, we have,

$$\left(-\sum_{i=1}^{E} \frac{1}{c} \frac{\partial}{\partial t^i} \frac{\partial}{\partial t^i} + \sum_{i=1}^{P} \frac{\partial}{\partial x^i} \frac{\partial}{\partial x^i} - \left(\frac{m_0 c}{\hbar} \right)^2 \right) \phi = 0 \qquad (3.25)$$

How is the six-dimensional Klein-Gordon equation related to the Schrödinger equation?

Theoretically, the six-dimensional Klein Gordon equation takes as argument a spatiotemporal field while the Schrödinger equation takes as argument a wave function on phase space. Occasionally, as a relativistic particle is carefully considered, the six-dimensional Schrödinger equation may be refined, in a relativistic way, into a six-dimensional Klein-Gordon equation. The state of a physical system is specified by the coordinates from each of the axes of a multidimensional space or phase space. Each unique point in the phase space has a possible corresponding state. All possible states of a system are represented in a phase space according to dynamical system theory.

§ 4. The Color Charge Field

The Gluon Standard Model includes a color charge field to endow particles their masses and break their electroweak symmetries. The color charge field permeates the spatiotemporal medium and can break the symmetry laws of the electroweak interaction causing the $W^\pm$ and Z^0 bosons to gain mass through the color field contraption at temperatures below an extreme high temperature. As the electroweak force bosons gain mass, these bosons are limited to travel through very short distances, as shown empirically. The color charge field is the primary mechanism to endow mass to elementary particles and composite particles. This color charge field is a scalar field that has a constant non-zero value in the spatiotemporal vacuum.

It is proposed that the color charge field may be considered the zero-point energy field of the spatiotemporal vacuum. After the Big Bang, this zero-point energy field gave the very early universe a symmetry that was very smooth, extremely high energy, uniform, and indistinctive. It is possible to theorize that consecutive breaks of symmetry at phase transitions of our universe may have allowed the emergence of our current four forces of nature and their related fields. Even though, the present zero-point energy field is very low, the expected energy from the color charge field, color supersymmetry, under the color string theory, would be of a similar magnitude. As previously discussed, the color string theory provides a potential solution to the cosmological constant problem, or the vacuum catastrophe, which is the variance between the theoretical large zero-point energy value suggested by current quantum field theory and the observed values of vacuum energy density, the small positive non-zero value of the cosmological constant, which is closely related to the missing fermionic mass.

4.1 The Unconfirmed Crucial Hurdle.

As technology develops, the empirical evidence that color charge field exists and couples to the Higgs field would be revealed since the predictions of a theory can lead researchers to believe that the theory is correct or incorrect based on its premises. A premise is a statement that an argument claims will induce or justify a conclusion; in this case the conclusion is the veracity of the Color String Theory and its related Color Supersymmetry. Furthermore, the affirmative conclusion would indicate that the Gluon Standard Model is correct and fundamental for particle physics.

It is possible to state that scientists have not found a way yet to determine if the color graviton field exists, because the technology needed for its detection has not been built yet, but that could change as technology develops rapidly. If the color graviton exists, then the color graviton field should also exist as a fundamental field, and the Gluon Standard Model is correct and genuinely fundamental. So, there needs to be an active and ample search for the color graviton to prove its existence as the existence of the color graviton remains an unconfirmed crucial part of the Gluon Standard Model.

4.2 The Quest for the Detection of the Color Graviton.

Even though the color graviton may exist everywhere in the spatiotemporal medium, finding a color graviton or a polymer of color gravitons is quite a challenge. As soon as there is a technique to detect a perturbation, a potential state, or an excitation, that may be manifested by a color graviton, or by color gravitons, with the correct energy requirement to produce color gravitons, and the extremely sensitive equipment to detect them, there would be a way in principle to find the empirical evidence. Particle colliders, detectors, and super computers are expanding their capabilities very rapidly to develop the equipment and techniques to accomplish this task.

Researchers at particle accelerator facilities around the world have been studying quarks, gluons, protons and neutrons inside the nucleus with giant, powerful microscopes, that enable scientists to see things a million times smaller than an atom. This unprecedented view of the basic building blocks of ordinary matter and its interactions allows researchers to gain deeper insight into the particles and forces in our physical reality. It would be only a matter of time for these advanced facilities to create color gravitons and other particles for observation and study.

The discovery of the color graviton would be the fifth force carrier particle to be discovered in nature. The color graviton is predicted to be a spin-2 force carrier because the source of gravitation is the stress–energy–momentum tensor. The color graviton mediates gravity, while the Higgs boson does not. The Higgs gives particles a rest mass, but it is not the only source of rest mass. The color graviton would have to be empirically confirmed to behave, interact, and decay as predicted by the Gluon Standard Model, as well as having the above fundamental attributes of a force carrier.

So, does the Higgs boson emit color gravitons? Color Gravitation couples with matter, including the Higgs field at a quantum level because the Higgs field is a mass field, so the color gravitation field couples to the Higgs field the same it couples to other gauge bosons, elementary particles, or systems of mass, through the gravitational radiation from the color gravitons. For instance, as intrinsic angular momentum or spin. In fact, *it is possible to postulate that any source of mass or energy may be a source of gravitation in the absence of other forces*, as that source obeys the law of inertia to follow a geodesic path determined by the gravitational field that may be present. Quantum gravitational radiation from color gravitons at the quantum energy limit will not modify the above postulate.

The Higgs field and the breaking of symmetry may give or increase the attribute of mass of elementary particles, including the ability to produce greater quantum gravitational radiation to some extent due to the spatiotemporal expansion or contraction, or through classical gravitation. Even though, the additional mass from Higgs field is predicted to be approximately one percent. The other ninety nine percent is attributed to the mass-energy content of the strong force of the Gluon Standard Model, unrelated to the Higgs scalar boson.

If and when the theoretical existence of the color graviton is confirmed, the concept of the color graviton field can be strongly supported. The presence of the color graviton field would explain, not the Higgs scalar boson, why and how it mediates gravity between elementary particles, The existence of the graviton field would also explicate why the very long range of gravitational force from each mass, or system of masses, in all spatiotemporal directions, but with the gravitational force reducing rapidly with distance.

§ 5. The Color Graviton Theory

What is a color graviton?

It is not known if a mass could emit color gravitons; the color graviton is a virtual boson that carries the gravitational force and its field effects between masses. Quantum gravity is related to a color graviton. In quantum field theory, the color graviton is massless with a spin of 2 and it moves at almost the speed of light of a photon. This is because the source of gravitation is the stress–energy–momentum tensor, a second-rank tensor. Aside, a tensor is an algebraic object that represents a real-valued multilinear function in each of its arguments, with an input or several inputs that produce an output in a coordinate system. The rank of the tensor is the number of inputs of the tensor to produce its output. A tensor, like a vector, may be invariant under coordinate system transformations. A tensor may represent a vector or a matrix of n-dimensions that represents all types of identical type of data. The element of a zero-dimensional array is a spatiotemporal point or a scalar.

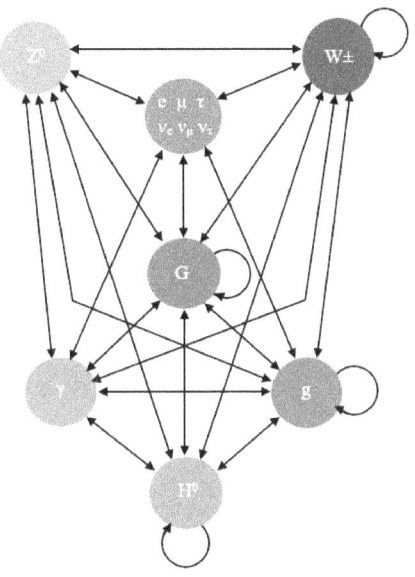

Figure 4. A diagram that illustrates the interactions between the graviton and fundamental particles in the Gluon Standard Model.

How is the color graviton related to gravity?

The color graviton may be defined as a nonlinear quantum of gravitational radiation. The Color Graviton Theory is a Quantum Gravity Field Theory. The mass of a particle, or a system of particles, creates a quantum field via the color gravitons. It is possible to quantize the Color Graviton Theory by looking at oscillations around the color graviton field background with probable solutions to the classical equations of motion, which are extremal action solutions that govern the path integral.

Consequently, when a particle is accelerated, its mass emits gravitational radiation to causally transmit the information of its motion to the rest of the masses that constitute the gravitational field. If this gravitational radiation is from a quantum source it can be quantized as an emission of real color gravitons, that carry real energy and linear and angular momenta. Hence, a color graviton is a quantum of gravitational radiation. The color graviton may be expressed as a gravitational spinor.

A color graviton carries a discrete amount of gravitational energy and an associated gravitational mass. The color gravitational potential energy consists of discrete amounts, or quanta, of color gravitons that are convertible to or from quanta of light, or electromagnetic energy quanta like photons.

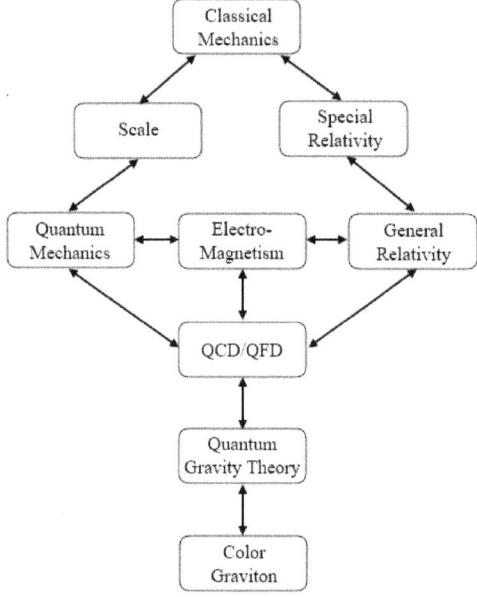

Figure 5. A Diagram of Current Classical and Quantum Theories.

The quantum of gravitational radiation of a color graviton was defined as

$$g_s \equiv \pm m'_s \omega_s^2 \ell_s^2 \qquad (5.1)$$

Where "$\pm \omega_s$" is the intrinsic angular frequency, "m'_s" is the relativistic mass, and "ℓ_s" is the length of a color string. The color graviton that travels in the retarded spatiotemporal wave is a relativistic luxon. Let us define the color graviton Planck constant as follows:

$$\hbar_s \equiv \frac{m_s \ell_s^2}{t_s} \qquad (5.2)$$

Where "$t_s = \ell_s/c$" is a characteristic temporal period for a color string where "ℓ_s" is an approximate characteristic length for a color string equal to 10^{-35} m, "m_s" is the infinitesimal mass of a color string, "ℓ_s" is the length of a color string, and "$\hbar_s$" is approximately equal to $5.827965384 \times 10^{-87}$ (Kg·m²)/s.

Thus, a quantum of gravitational radiation of a color graviton may also be defined as

$$g_s \equiv \pm \hbar_s \omega_s \qquad (5.3)$$

According to Color String Theory, color gravitation is an exchange mechanism among the masses of a system. If the interaction between the nucleus of an isolated hydrogen atom and its electron, a color graviton may travel toward the nucleus as another color graviton may travel to the electron, both color gravitons attract each mass toward the other by a corresponding quantum of gravitational radiation. However, it is theorized that the gravitational waves of the color gravitons interfere constructively or destructively in the spatiotemporal background, and a resultant gravitational wave may or may not emerge. Hence, in this isolated case, the nucleus and the electron may be affected by resultant gravitational radiation.

Can color gravitons escape the event horizon of a supermassive black hole?

Color Gravitons are massless gravitational gauge bosons, like photons. Color Gravitons or photons follow null geodesic paths. Thus, neither gravitons nor photons can escape the event horizon of a supermassive black hole. Consequently, the fact that the gravitational radiation of a supermassive black hole may be detected even when color gravitons cannot escape the event horizon of a supermassive black hole explicates that a color graviton quantum gravitational field is not unique, and that gravitational waves may be radiated in several ways using different mechanisms to create a spatiotemporal perturbation or wave. If the color graviton were the only way to mediate gravity, the above observation would not be correct. Nonetheless, the color graviton is a quantum gravitational mechanism for color string theory. It is well known that gravitation from a celestial body is nonlinear, and from previous research, it has been theorized that gravitational waves are radiated in a six-dimensional spatiotemporal medium.

Therefore, the color graviton, as a gauge boson, can describe a quantum of nonlinear gravitational radiation as a Quantum Gravity Field Theory such as the Color Graviton Theory. Gravitational waves may define gravitational radiation in a quantum gravity field with intrinsic angular momenta (spins) and energy without any direct involvement of color charges. Thus, gravitational waves do not have to interact with matter that they encounter as they propagate through the spatiotemporal medium. But as they propagate through matter, or near a mass or system of masses, they can interfere constructively or destructively, with the gravitational radiation from the kinetic motion of the particles of matter or with the color graviton gravitational radiation. The quantum gravitational interference with the constituents particles of matter present a quantum gravitational perturbation within matter.

Hence, it is reasonable to theorize that color gravitons may have intrinsic angular momentum and energy as a color string loop, a color charge polymer, or a color charge pair, that can interact with the constituents of matter at a quantum scale. Researchers in Quantum Mechanics may find spin to be quite mysterious. Spin in Quantum Mechanics is associated to the four main types of symmetry: translation, rotation, reflection (mirror), and glide reflection.

So, the following explanation of a spin-2 color graviton is a metaphor and only a visual aid. Let us imagine a Cartesian coordinate system with the +z-axis upward, the +x-axis to the right, and the +y-axis that is out of the page. A spin-2 particle is going to rotate on only two axes, the +z-axis and the +x-axis, while the +y-axis is the axis of gyration.

The Color Graviton is a spin-2 boson with two symmetry axes, reflection symmetry, and ninety degree anti-symmetry. If the color graviton turns to a right angle it interchanges its +z-axis and its +x-axis, so the +z-axis keeps the clockwise motion and the +x-axis changes to a counter-clockwise motion, reversing the orientation. Two rotational interactions of the spin-2 color graviton are non-zero, but they are symmetric. So, if the z-x plane is flipped over the color graviton system is returned to the same spin-rotation interactions. The Color Graviton would have the same spin-rotation interactions under 180^0, 360^0, 540^0, and 720^0 rotation.

Hence, if the Color Graviton spins clockwise around the +z-axis and the +x-axis, and then turns 180^0 clockwise around the perpendicular +y-axis to an upside down position, the side facing down around the z-axis would turn counter-clockwise. Moreover, the clockwise rotations of the +z-axis and +x-axis would be reversed, while also reversing both axes, as the color graviton returns to its original state of spin-rotation interactions.

Figure 6. Illustration of Spin 2 of a Gauge Boson to its Original State.

So, this quantum interaction may be in the form of the intrinsic angular momentum, energy, and wave interference, with the inclusion of color charge interaction, a sort of chromogravitonic quantum interaction. As the color gravitons spin they generate gravitational radiation that may offset external gravitational fields to some extent, through a counter gravitational effect.

Thus, the color graviton may be considered a gauge boson, or a quantum particle-wave for gravitation, or a force carrier of the Gluon Standard Model that has a predicted long Compton wavelength of

~1.6×10^{16} *meters*, a very small frequency of up to 1 KHz max, and a very long field range. Unlike other gravitational fields, the color graviton field may interact with or influence other waveforms such as electromagnetic waveforms.

Aside, in general, gravitational waves are expected to have frequencies of 10^{-16} Hz $< f < 10^4$ Hz, while gravitational waves affect light, all light in the presence of a gravitational field may bend or shift its frequency.

Let us imagine that two free color gravitons moving very fast while spiraling inward or outward, very slowly around a common point in a color charge field, while not being able to attach to one another, can develop quantum gravitational waves, or spatiotemporal waves that are moving outward or inward, and may interfere with other spatiotemporal waves around them, while the center spatiotemporal medium around the point is expanding or contracting independently from the space-time beyond the spiraling orbit of the color gravitons.

A similar physical phenomenon happens in the large-scale of classical astrophysics, a binary star system loses angular momentum as the two orbiting stars spiral towards each other, the angular momentum is radiated away by gravitational waves.

The inward or outward spiraling motion of the color gravitons depends on the attractive or repulsive color charge force between the particles as they approach each other.

The closed loop topology of the color gravitons may avoid the attachment between the particles and provides viable time for the rotation of the pair to emanate a significant gravitational field that manifests the emergent properties of General Relativity from space-time.

Hence, it is possible to theorize that each of these gravitational waves is an instant of what may be considered a quantum of gravitational radiation of a color graviton pair, "g_1g_2".

This image of the graviton is a spatiotemporal perturbation of the medium by two gauge bosons. So, in this case, the color graviton pair would have all the properties of quantum gravitational waves.

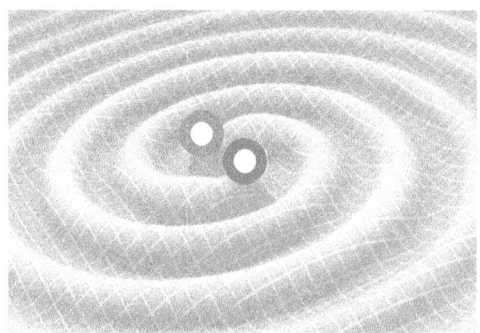

Figure 7. Two Color Gravitons Spiraling in the Spatiotemporal Medium.

Could the color graviton be considered a particle-wave manifestation? For if that is the case, gravitation may be quantized as a particle and a wave. It is theorized that a gravitational spinor is an instant of quantized angular gravitational momentum or spin that may be formulated as an interlaced spatiotemporal fabric where General Relativity emerges as a classical theory. Nonetheless, the emergence of a gravitational field from the expansion or contraction of space-time at any arbitrary spatiotemporal point from the constructive or destructive interference of gravitational waves, or around a body of mass, does not imply angular gravitational momentum or spin since a particle or a pair of particles may not be necessarily involved.

Consequently, there is more than one way for gravitation to emerge in our universe. Gravitational waves may expand, contract, spin, or be static, depending on the gravitational source or sink, and other spatiotemporal perturbations that may be present in the spatiotemporal medium.

Our familiar understanding of gravity, which is that gravity is an attractive force, is not complete. Gravity may be attractive or repulsive, counter-gravitational (destructive interference) or pro-gravitational (constructive interference), but not simply gravitational, even though that is the way we typically think of it for obvious reasons.

§ 6. The Cosmological Graviton.

The cosmological graviton "g_Λ" may be defined as linear quanta of gravitational radiation between two arbitrary spatiotemporal points of the medium. It would be advantageous to define the cosmological graviton or anti-graviton in terms of Planck units.

$$g_\Lambda \equiv \pm m'_\Lambda \omega_P^2 \ell_P^2 \qquad (6.1)$$

Where "ℓ_P" is the resultant Planck wavelength of the cosmological graviton, "ω_P" is the Planck frequency of the cosmological graviton, which is equal to $\omega_P = 1/t_P$, and "m'_Λ" is the relativistic mass.

Gravitational waves are able to carry energy and momenta and by doing so they carry those away from the source. The cosmological graviton would be a massless spin-2 boson because its gravitational force appears to have unlimited range. A massless spin-2 boson field would give rise to a gravitational force, because the field must interact with, or couple with, the stress–energy–momentum tensor in the same way that another gravitational field does. The cosmological graviton has the potential to unite quantum theory with gravity.

The cosmological graviton is a virtual particle-wave with a relativistic virtual mass given by

$$\left(m'_\Lambda\right)^2 c^4 \equiv E_G^2 - p^2 c^2 \qquad (6.2)$$

$$m'_\Lambda \equiv \sqrt{\frac{E_G^2 - p^2 c^2}{c^4}} \qquad (6.3)$$

If there is a difference in the energy and momenta of the cosmological graviton, when the angular frequency is conserved and the linear frequency increases during compression of its wavelength, the cosmological graviton would gain relativistic mass or virtual mass. As the virtual cosmological graviton gains relativistic mass, the spatiotemporal curvature about the graviton increases.

The rest mass "m_0" of the cosmological graviton, as very near the event horizon of a black hole due to time dilation, in terms of an infinitesimal cosmological constant "Λ", may be defined as

$$m_0 \equiv \frac{\hbar}{c}\sqrt{\frac{2}{3}\Lambda} \qquad (6.4)$$

Let us consider the following postulate (Eddington's Postulate): Every point in space-time expands freely in all directions unless obstructed. We may hypothesize that this postulate is supported by the Huygens Principle. (Nieves, 2020)

Every point on a local spatiotemporal wave-front in isotropic homogeneous space-time may be considered a source of secondary spherical spatiotemporal wavelets which spread in the outward direction at the speed of time (light). The new spatiotemporal wave-front is the tangential surface to all, or of all, of these secondary spatiotemporal wavelets.

The principle that any point on a spatiotemporal wave-front may be regarded as the source of secondary spatiotemporal wavelets and that the surface that is tangent to the secondary spatiotemporal wavelets, the envelope, can be used to determine the future position of the spatiotemporal wave-front supports Eddington's Postulate.

If we consider an extended line of space-time points, the resultant spatiotemporal wave will consist of an infinite number of space-time points and may be thought of as generating a plane spatiotemporal wave front. If a space-time locality is isotropic and homogeneous, allowing time to expand with the same speed regardless of its direction of propagation, the three-dimensional spatiotemporal envelope of a space-time point will be spherical.

Complex and interacting spatiotemporal points

In addition to previous research, the emergence of classical geometry may begin with the quantum entanglement between adjacent points in the spatiotemporal medium. As a spatial manifold expands, contracts, or remains static, the quantum entanglement between adjacent points may increase, decrease, or remain the same. This is the basis for spatiotemporal thermodynamics. The quantum entanglement between points is the framework of causation, or cause and effect, between events in physical reality.

An arbitrary point in space exists in the physical reality of the spatiotemporal medium that may be expressed as a real number. A real or complex point endows one or more real or complex points that are adjacent and entangled. It is possible to suggest that since space-time is complex, a spatiotemporal point may be represented by a real, an imaginary, or a complex number. An imaginary (temporal) point may precede its spatial (real) point, depending on the direction of the arrow of time. All spatial points that exist that are embedded on a manifold may be regarded as real, or as real source points, preceding points on a manifold may be considered imaginary, and proceeding points may be considered imaginary. Points that are in a network state may be interacting in a fundamental and emergent six-dimensional space-time that underlies the applicable conservation laws.

The entanglement between adjacent points may be represented as causal networks that exists in our universe. If two points interact, they are adjacent or temporally related. So, if two interacting points are adjacent, the adjacency matrix is $A_{ij} = 1$, and if they are not, $A_{ij} = 0$. Two future interacting points may or may not be adjacent, 1 or 0. Hence, point entanglement may be considered a quantum bit, $A_{ij} = \{1, 0, X\}$, such as the state $|\psi\rangle = \alpha|0\rangle + \beta|1\rangle$. The quantum property of entanglement may lead through network analysis to the underlying framework of an emergent and fundamental spatiotemporal background, to describe the emergence and evolution of classical geometry.

It is possible to theorize that the entangled complex numbers may represent complex points on three adjacent layers of past, present, and future spatiotemporal surfaces in a six-dimensional coordinate system with one spatiotemporal dimension passing through the origin of each layer. The operator "$+i$" may be utilized to go from a real present point to a future imaginary point or from the past imaginary point to a real present point. The operator "$-i$" could be utilized in the opposite direction. The spatial flow from the past to the future may be opposite to the temporal flow from the future to the past at the same spatiotemporal locality as previously theorized in "A Dynamic Theory of Space-Time". As space or time expands or contracts through a dimension, they would follow the retarded or the advanced direction of the wave function at an arbitrary point.

Complex number analysis may be an effective tool to demonstrate the shifting of spatiotemporal points during expansion or contraction of the medium as time passes. The flow of the interaction between points may be bi-directional and multidimensional in space-time. Hence, it is possible to suggest that points or objects that may be non-local or separated in space, may be local in time. A de Broglie stochastic process that underlies quantum mechanics can explain how a particle would hop from a point on a spatiotemporal wave to a point on another. (de Broglie, 1967)

From previous research, the principle of superposition of spatiotemporal waves may be stated as follows:

When two spatiotemporal wavelets interfere, the resulting displacement of space-time at any locality is the algebraic sum of the displacement of the individual spatiotemporal wavelets at the same locality in space-time. (Nieves, 2020)

For instance, if two spatiotemporal wavelets have a displacement in the same direction at any locality along space-time, constructive interference will occur between the spatiotemporal wavelets and a color anti-graviton would be produced, $-g_\Lambda$. If two spatiotemporal wavelets have a displacement in the opposite direction at any locality along space-time, destructive interference will occur between the spatiotemporal wavelets and a color graviton would be produced, $+g_\Lambda$. Therefore, the cosmological graviton "$\pm g_\Lambda$" is directly related to the spatiotemporal expansion or contraction of the medium of our physical reality.

How does the color graviton exchange mechanism work?

Gravitational waves have been detected, so do they exhibit particle-wave duality as well? When will the graviton be detected? Photons are known to exhibit particle-wave duality and quantum properties. Will the graviton also exhibit those particle-wave properties? Is there a particle like the predicted graviton in gravitational radiation that is distributed in quanta?

So far, we have discussed three gravitational mechanisms theorized to exist in our physical reality, namely, the classical gravitation due to spatiotemporal geometry generated by the spatiotemporal divergence or convergence, the gravitational radiation by a system of mass, and the cosmological graviton. The color graviton exchange would be a fourth gravitational mechanism that we can discuss, without implying that there would not be any other gravitational mechanisms to be found in our physical reality.

The General Theory of Relativity described gravitation as the curvature of the spatiotemporal medium, but researchers have long sought a theory of quantum gravity, and a gauge boson for gravitation like the color graviton. It is theorized that even though the color graviton has not been detected yet, the mediating particles for gravitation roam the quantum realm in droves, and their intrinsic property and interaction gives rise to the classical gravitational force. This enduring expectation for quantized gravity has to make mathematical sense under close examination, after making precise calculations on possible graviton interactions that do not result in an answer of infinity.

With respect to the exchange mechanism of the color graviton, let us describe a definition of the color graviton. The color graviton is not necessarily spherical as it is typical for an elementary particle, instead it is a quantum gravitational loop made from a color string. Color gravitons are around a mass or system of masses in layers of density that represent the layers of the contraction of spatial dimensions and the expansion of the temporal dimensions. The color gravitons exist in gravitational waves that radiate from the particle of mass or from a system of mass. Thus, a color graviton may be considered a particle-wave.

The color graviton carries its unique color force and gravitation due to its infinitesimal mass, and it is adhesive to other color gravitons that it is attracted to by its own color force. Therefore, as gravitons glue together they do not attach, but they may form polymers of gravitons that have larger mass with a different spin other than a spin of 2. Gravitons can be exchanged between masses as both gravitational radiation or graviton particle that interacts with the property of mass and its gravitational field. Hence, the graviton has its inherent quantum gravitational field and it can also be a part of the stronger gravitational radiation of a particle field or the field of a system of particles. This view of the color graviton as the gravitational stowaway, or clandestine gravitational rider, provides a particle-wave concept for quantized gravity where the wave is a spatiotemporal carrier in the same way that it is for a photon as proposed in "A Dynamic Theory of Space-Time". (Nieves, 2020) Every spatiotemporal point expands or contracts in all directions, allowing the gravitational radiation and the exchange of gravitons among the objects of mass during the expansion of space-time.

Could gravitons be detected?

After the declaration that gravitational waves were detected, there is renewed interest in the existence of gravitons. There is an analogous correspondence between photons and electromagnetic waves and gravitons and gravitational waves that has intrigued researchers of the gravitation force. Both the existence of the photon and the graviton have been deduced before their actual detection. Since experiments with low intensity light were effective to detect photons, are there experiments with low intensity gravitational waves that could be effective to detect gravitons? The gravitons are expected to emerge if gravitation is fundamentally quantized in our physical reality. The graviton is expected to generate quantum fluctuations as its spatiotemporal medium expands or contracts.

Some of the properties of gravitational waves are

- Gravitational waves expand and contract in orthogonal directions as they travel through the spatiotemporal medium while exhibiting wave properties.

- Gravitational waves interfere constructively or destructively with other gravitational waves or spatiotemporal disturbances in their medium.

- Gravitational waves carry measurable quanta of energy that could be detected.

- The wavelengths of gravitational waves are extended or contracted, losing or gaining energy, as they travel through their expanding or contracting spatiotemporal medium.

- Gravitational waves propagate at a specific speed that is slightly less than the speed of light, approximately by not more than one part in 9.85252×10^{14} or $\sim 10^{15}$.

Let us consider an approximate but useful analogy to the gravitons and their gravitational waves for visualization that involves the waves of the ocean and an extremely large number of inflated ring-shaped life preservers, or lifesavers, that float freely on the surface of the water, as each lifesaver orbits around an arbitrary point on the ocean waves.

The lifesavers approach each other as they orbit around or travel, but could bounce off each other, assisting the actions of the waves as a framework that continuously expands or contracts. Each lifesaver would also move up and down, back and forth, along the undulating surface of the water, as the molecules that make up the water may move in a similar fashion.

A framework of gravitons moving along circular paths can appear to create a large-scale vision of gravitational waves.

Similarly, individual lifesavers that move in a distinct order may create large-scale lifesaver waves that resemble ocean waves, and the gravitational waves that can be measured are theorized to consist of distinctive quantum particles like gravitons that may produce wave patterns that assist their fundamental gravitational wave.

Color Gravitons can assist the spatiotemporal undulation between the spatiotemporal source nodes if they orbit around an arbitrary spatiotemporal point of expansion or contraction, as they attract or repel the adjacent color gravitons that are also orbiting nearby, to create and follow the wave characteristics of gravitational radiation in all spatiotemporal directions.

This self-sustained wave mechanism may serve as a long range propagator, or a graviton wave drive, as well as a grid substructure for the reinforcement of the gravitational waves through the spatiotemporal medium as color gravitons stowaway on the waves emitted by a particle mass or a system of particles.

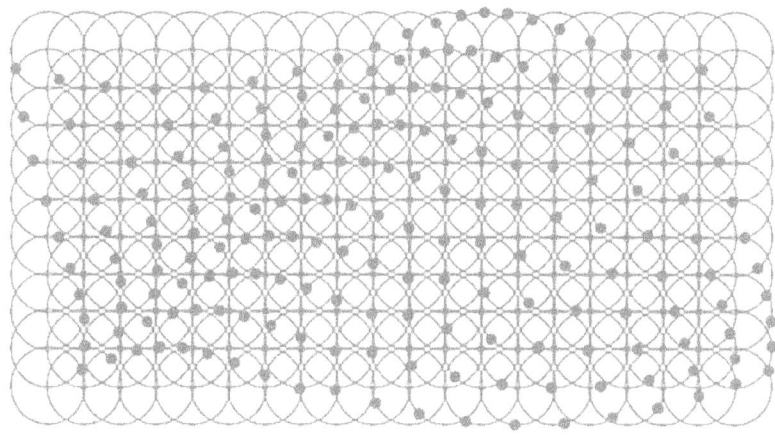

Figure 8. An Illustration of a Discrete Waveform.

The color graviton, a gauge boson, is theorized to have an inherent infinitesimal mass and a color charge, like other bosons such as the Z^0 or $W^{\pm}$ bosons of the weak interactions that have mass, and in the case of the $W^{\pm}$ boson, it has electromagnetic charge.

Thus, the inherent infinitesimal mass of the color graviton has not been excluded by precise calculations. The quest to design an effective graviton detector that avoids an enormous circular particle accelerator design is crucial to detect gravitons.

The possible quantum effects of gravitons:

- Each graviton mass may have a signature. Could the signatures of gravitons be detected at the shortest distance of a quantum scale where the gravitational field would be strongest and the quantum effect of gravitons are most pronounced? For example, it may be possible to conduct a virtual experiment as close to two naked singularities at the exact moment when they are merging, or two black hole singularities that are merging, in a computer-simulated spatiotemporal background where phenomena inherent to quantum gravitation may emerge at very rapid timeframes. Moreover, it might be possible to use multiple interferometers and very high frequency laser pulses ($\sim 10^{-18}$ s) available to detect signatures of gravitons.

- Gravitons may have an inherent orbital frequency and wavelength. What frequency and wavelength of gravitons reproduce quantum effects on a fundamental gravitational wave?

- A graviton mass travels slower than light while gravitational radiation can travel at the speed of light. What harmonics in the recorded signal of a fundamental gravitational wave have slower speed than light by a factor no greater than one part in 10^{15}? Could a Fourier linear frequency analysis reveal the velocities of these harmonics?

- Gravitons may have inherent quantum harmonic waves. Could the amplitude of the quantum harmonic wave of a graviton be detected?

An ultrashort laser pulse may be amplified by a chirped pulse amplification technique up to the petawatt level with a laser pulse that is first extended, then amplified, and contracted again both spectrally and temporally through the apparatus. The different color components of the ultrashort laser pulse would travel at different distances. The chirped pulse amplification technique can be used by the most powerful lasers today.

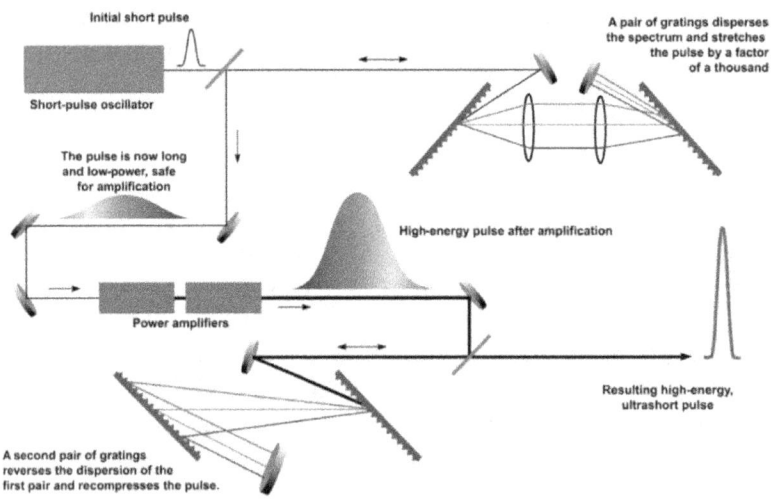

Figure 9. Chirped Pulse Amplification. (Lawrence, 1995)

The majority of the signatures that may be detected that would demonstrate if there were quantum gravitation would not necessarily reveal the existence of gravitons.

Nonetheless, if B-modes were detected as predicted that would infer that there were inherent quantum gravitation, even though it would not detect either the presence of gravitons. B-modes are a pattern of polarized light originating from the inflation of the Big Bang.

Similarly, a double slit experiment involving electrons to detect if their gravitation passed through either slit, or through both slits, and if there were quantum gravitation, still would not directly detect the presence of gravitons.

However, there are other potential signatures that may reveal the existence of gravitons. For example, if there were particles that were in quantum state superposition that depended on gravitational self-energy levels, it may be possible to detect if there were quantum gravitation.

Or if photons of various wavelengths passing through a crystal made the crystal move in steps instead of continuously, then it could be inferred that there may be quantized space.

The quest of directly detecting gravitons would lead to a great achievement that requires advanced technology for the right experiment. Photons, gravitational waves, and gluons, travel at the speed of light, to carry the electromagnetic, the gravitational, and the strong nuclear interaction. If the graviton exists as a particle with non-zero mass, then it would not be a luxon and it would be measurable.

It is possible to hypothesize that since gravitational waves have been detected from a signal that came from a black hole merger, the existence of the graviton becomes more certain, as General Relativity maintains its validity in the weak field limit of gravitation. Thus, as perturbations in the spatiotemporal metric propagate as gravitational waves, the existence of gravitons is implied by Quantum Mechanics. The detection of a graviton would be a different matter if a graviton interacts weakly with matter according to present theory. A different strategy to find a graviton would be to observe the cosmic microwave background because gravitons would oscillate with very short wavelengths that have acute fluctuations. According to the cosmological model of inflation, these very short wavelengths would have stretched to longer extents through the cosmos. So, quantum gravitation should be observable as swirls in the alignment or in the polarization of photons in the cosmic microwave background.

Nonetheless, the B-modes of these intense swirls depend on the time and energy of cosmic inflation. The measurement of the B-modes may provide the properties that agree with the present cosmic inflationary theory that would be strong evidence of the existence of quantum gravitation. Hence, by looking directly for acute fluctuations in gravitational waves which are believed to consist of gravitons from the very early universe, it may be possible to find quantum gravitation. A gravitational wave observatory, or laser interferometer space antenna, in space, that has very sensitive equipment to detect oscillating gravitational waves, may do the job.

§ 7. The Jiggle of a Quantum Gravitational Wave.

A gravitational wave is a spatial wave that ripples through the spatiotemporal medium that may or may not endow more space as every spatiotemporal point on the wave may expand, contract, or be static. The gravitational wave may be considered the pilot wave of gravitons and other particles of mass, massless particles, or virtual particles. (de Broglie, 1927, and Bohm, 1952)

The guiding equation for the phase velocity of the orbiting gravitons around a spinless arbitrary point of curvature on the surface of a gravitational wave, the phase velocities of the orbiting gravitons are given by

$$\frac{d\rho_k}{dt}(t) = v_k r_k \operatorname{Re}\left(\frac{\nabla_k \phi}{\phi}\right)(\rho_1,...\rho_N,t) \qquad (7.1)$$

For an orbiting graviton, the phase velocity can be written as

$$v_k = \frac{2\pi\lambda_k}{T_k} \qquad (7.2)$$

Where "T_k" is the period, "ρ_k" is the position of a graviton, and "λ_k" is the wavelength of the phase velocity, v_k. The areal velocity of any orbit is constant, a reflection of the conservation of angular momentum, but not the phase velocity of the expanding and contracting framework along the path of the gravitational wave.

Where the variations of the phase velocities "v_k" correspond to the positions of an "N" number of orbiting gravitons, while "ϕ" represents a real-valued function on the spatial gravitational wave. In the case of a spatial gravitational wave, the influence of all of those orbiting gravitons can be enfolded into an effective framework wave function for a subsystem of the spatiotemporal medium. Gravitons carry the gravitational force in a manner similar to how photons carry the electromagnetic force. Gravitational waves are thought as a collection of gravitons, in the same way that a collection of photons are visualized as light rays. So, a collection of gravitons may transform a gravitational wave.

As a gravitational wave passes through a gravitational wave detector, the distance between the end sensors of the detector that behave as if they were two masses, will extend or contract, as the wave modulates the distance between the ends. Since gravitons theoretically ride on the gravitational waves, that act like pilot waves for gravitons, when detector sensors, or masses, absorb or emit gravitons, the masses jiggle randomly as graviton noise. The stronger the wave, the greater the jiggle, and the more straightforward may be the detection.

Contingent upon how a gravitational wave is produced, the gravitational wave may have different quantum states. Generally, a gravitational wave is generated in a coherent state. A jiggle on the spatiotemporal medium. A gravitational detector must be tuned to these usual gravitational waves that may be emitted from celestial collisions of spiraling heavy neutron stars or supermassive black holes.

As the gravitational tidal field of a source changes with time, those changes propagate out from the gravitational source at the speed of light, "c". These changing tidal fields constitute gravitational radiation. If the changes are continuous or oscillatory, they become gravitational waves.

The amplitude of the gravitational signal or the dimensionless strain "h" of the gravitational signal for a spiraling binary neutron star pair. Hence, the strain "h" is given in square radians, demarcating the spatiotemporal envelope of the gravitational wave.

Therefore, "h" is twice the fractional change in gravitation over a displacement between two nearby masses due to the gravitational wave, $h = 2\iint \frac{\Delta g}{d} dt^2$. This change in displacement occurs in the plane transverse to the direction of the gravitational radiation, and causes an expansion along one axis and a contraction along the orthogonal axis. The net spatiotemporal distortion is twice as much as an expansion or a contraction alone, which is the reason for the factor of 2 in the equations for "h". Two times the change in displacement "Δd" over the displacement "d". However, the strain "h" is not itself directly observable. A constant "h", or an "h" that varies linearly with time, is exactly equivalent to starting the masses with a small relative velocity or at marginally different positions. The gravitational radiation would only be demonstrated by a second or a higher derivative of "h".

$$h \approx \frac{4\pi^2 GM (R_1 \cdot R_2)^2 f_{orb}^2}{c^4 r} \quad (7.3)$$

$$\sqrt{h} \approx 2\pi \left(\frac{v^2}{c^2} \right) \quad (7.4)$$

Where "v/c" is the ratio of the speeds of masses in the system to the speed of light "c", which would likely be much less than one, "R_χ" represents the radius of each neutron star, and "r" is the distance between the neutron stars. The units of "h" are in square radians, and the units of "$\sqrt{h}$" are in radians.

The graviton noise produced by the collisions of coherent gravitational waves is very little and difficult to measure. The jiggle produced by the gravitational wave detector as it absorbs gravitons from a gravitational signal is splendidly balanced with the jiggle the detector generates when it emits gravitons, which makes the detection of the graviton noise very difficult to perform. Nonetheless, there is a quantum state of gravitational waves that is "a contracted state" that generates a more measurable graviton noise. This contracted state increases exponentially as a collection of gravitons are contracted. Hence, the technique to measure these contracted states of graviton noise may be measurable after all, to obtain more information on the sources of these types of contracted gravitational waves.

Therefore, it is possible to theorize since light waves ride on temporal waves as well as contracted gravitational waves, there might be a wave to measure temporal deceleration within and around contracted gravitational waves since gravitational acceleration is the product of a spatiotemporal differential. Hence, the velocity of a temporal wave, the guiding spatial curvature, the principle of inertia of a moving body, and gravitational acceleration, are all the result of the spatiotemporal differential principle. Thus, a temporal wave does not accelerate, or decelerate, due to gravitational acceleration; a temporal wave changes its velocity due to a spatiotemporal differential across the spatiotemporal wave. The actions of the temporal field in spacetime provide measurable cosmological pressure at every point. Some of the massive and rapidly changing collisions mentioned above may produce contracted gravitational wave signals that are measurable graviton noise. The graviton signal must be discernible from any other signal that may be measured by the gravitational wave detector. So, the detector may have to be modified and tuned to the inherent properties of the graviton noise signal. (Nieves, 2020)

Chapter 6

The Lifetimes of Particles

§ 1. Why does a muon have a slower decay than an electron?

What is the acceleration of time around each particle? Is there more time dilation around a muon than an electron? Is there something we do not understand yet about particle physics? Does our current understanding of the universe account for every particle and force within it?

Are the calculations for such as strong interaction in a low energy regime, a non-perturbative regime, so precise that we are sure that it implies that the result may be new physics?

Is a major piece of the current Standard Model missing? The muon is just one flavor of leptons. A Muon is about 200 times heavier than an electron. A Muon has a negative charge and a spin. When the muon internal magnet is exposed to a strong external magnetic field of a particle accelerator the muon starts to wobble. The rate of this wobble is the g–factor or magnetic moment.

These technologies are merging to measure the anomalous magnetic moment "a_μ".

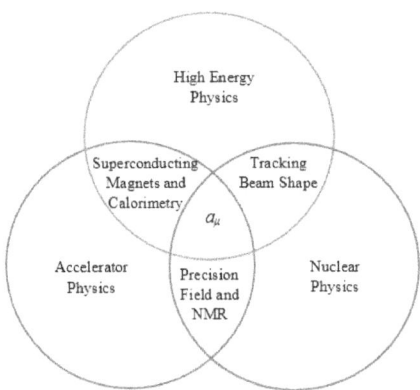

Figure 1. The Measurement of the Anomalous Magnetic Moment "a_μ".

§ 2. Is it possible that the g–factor has been hiding in plain sight all along?

The Landé g–factor (also called the "g" value or dimensionless magnetic moment) is a dimensionless quantity that characterizes the magnetic moment and angular momentum of an atom, a particle, or a nucleus. It is essentially a proportionality constant that relates the observed magnetic moment "μ" of a particle to its angular momentum quantum number and a unit of magnetic moment (to make it dimensionless), usually the Bohr magneton or nuclear magneton. The fine structure constant is denoted by the Greek letter alpha "α". A fundamental physical constant which quantifies the strength of the electromagnetic interaction between elementary charged particles. From previous research, "α" may also be described as the permittivity of the spatiotemporal medium.

$$\alpha = \frac{1}{4\pi\varepsilon_0} \frac{e^2}{\hbar c} \approx \frac{1}{137} \approx 0.007 \tag{2.1}$$

$$\frac{\alpha}{2\pi} \approx 0.001161715 \tag{2.2}$$

Quantum Electrodynamics (QED) can be described in technical terms as a perturbation theory of the electromagnetic quantum vacuum, or as higher order corrections of a perturbative series of the fine structure constant "α". As predicted by the eminent physicist Julian Schwinger in 1948, the radiative correction, $\alpha/2\pi$, to the magnetic interaction energy corresponds to an additional magnetic moment associated with the electron spin.

$$g_e \approx 2 + \frac{e^2}{2\pi\hbar c} \approx 2.001162 \tag{2.3}$$

In QED, radiative corrections change "g" from its Dirac value of 2. Corrections that may be symbolically expressed as Feynman diagrams.

$$\frac{g}{2} = 1 + C_1\left(\frac{\alpha}{\pi}\right) + C_2\left(\frac{\alpha}{\pi}\right)^2 + C_3\left(\frac{\alpha}{\pi}\right)^3 + C_4\left(\frac{\alpha}{\pi}\right)^4 + C_5\left(\frac{\alpha}{\pi}\right)^5 + \ldots + \alpha_{hadronic} + \alpha_{weak} \tag{2.4}$$

where the C_i are the coefficients in the QED contribution.

$$\frac{g}{2} \approx 1 + C_1\left(\frac{\alpha}{\pi}\right) + C_2\left(\frac{\alpha}{\pi}\right)^2 + ... \qquad (2.5)$$

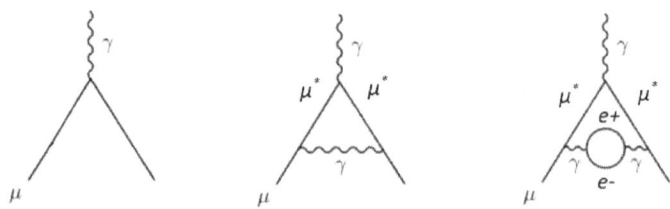

Dirac-Stern Gerlach   Schwinger Kusch-Foley   Vacuum Polarization

Figure 2. The Feynman graphs for the Dirac-Stern Gerlach "$g = 2 + (\alpha/2\pi)$", the Schwinger Kush-Foley "$+ C_2 (\alpha/2\pi)^2$", the lowest-order radiative correction first calculated by Schwinger, and the vacuum polarization contribution, which is an example of the next-order term. The " * " in μ^* emphasizes that, in the loop, the muon is off-shell.

The Stern-Gerlach experiment confirmed the quantization of electron spin into two orientations. Spin can take only two orientations, $\pm(1/2)\hbar$. There are two possible values for the S_z–axis, corresponding to the two spots on the observation screen in the direction of the magnetic field, taken to be in the "z" direction, as required by the fact that $s = \frac{1}{2}$ for electrons, i.e. they are spin–½ particles. The effective radius of the dotted circle is $\sqrt{3}/2\ \hbar$ or about $0.866025404 \cdot \hbar$.

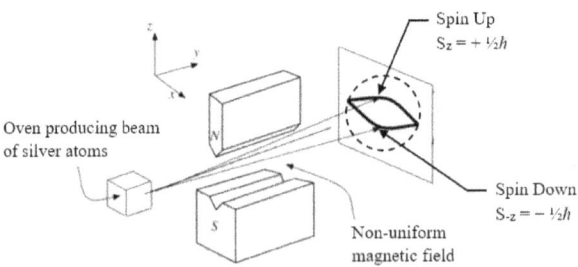

Figure 3. The Stern-Gerlach apparatus.

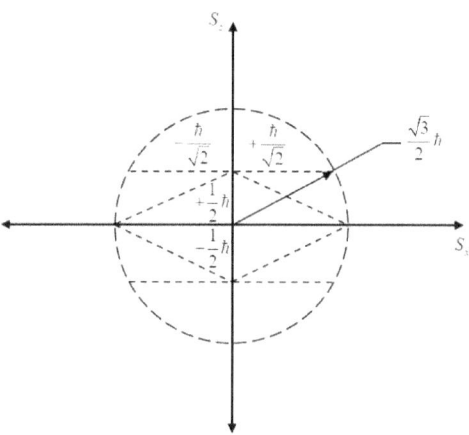

Figure 4. The Spin Orientations on the Observation Screen.

As the electron spins around, the axes of the precession of the electrons in the beam has an angular spin as the spatiotemporal medium expands in the directions of the S_x–axis and S_z–axis, projecting a radius of $\left(\sqrt{3}/2\right)\hbar$ as the particles hit the observation screen.

It is possible to suggest that the electrons are found to possess spin angular momentum "α_e" equal to $\left(\sqrt{3}/2\right)\hbar$ with a magnitude of $\sqrt{3}/2$, which requires the geodesic distances between two points on the spatiotemporal manifold over the particle to be expanding radiatively at a speed greater than "c", assisted by the angular spatiotemporal expansion. Hence, the spin of the electron may be describable in terms of the concept of the wavefunction of a particle. The effective value, $\pm\hbar/\sqrt{2}$, is the effective spin momentum "S_{eff}" parallel to the S_x–axis, where $\hbar$ is $m_p c l_p$ in Plank units.

$$\left(S_z\right)^2 + \left(S_{eff}\right)^2 = \left(S_{\alpha_e}\right)^2 \tag{2.6}$$

$$\left(m_e v_z z\right)^2 + \left(m_e v_x x\right)^2 = \left(m_e v_r r\right)^2 \tag{2.7}$$

Dividing by $m_e c d$, where "d" is a spatial distance, we have

$$\left(\frac{m_e v_z z}{m_e cz}\right)^2 + \left(\frac{m_e v_x x}{m_e cx}\right)^2 = \left(\frac{m_e v_r r}{m_e cr}\right)^2 \qquad (2.8)$$

$$\left(+\frac{\hbar}{2}\right)^2 + \left(+\frac{\hbar}{\sqrt{2}}\right)^2 = \left(\frac{\sqrt{3}}{2}\hbar\right)^2 \qquad (2.9)$$

$$\frac{\hbar^2}{4} + \frac{\hbar^2}{2} = \frac{3}{4}\hbar^2 \qquad (2.10)$$

$$\left(\frac{m_e v_r r}{m_e cr}\right)^2 = \left(\frac{v_r}{c}\right)^2 \qquad (2.11)$$

$$\left(\frac{\sqrt{3}/2}{1}\right)^2 = \left(\left|\frac{v_r}{c}\right|\right)^2 \qquad (2.12)$$

$$\left|\frac{v_r}{c}\right| = \frac{\sqrt{3}}{2} \approx 0.866025404 \qquad (2.13)$$

$$g_e - 2 \approx \frac{\alpha}{2\pi} \qquad (2.14)$$

$$g_e \approx 2 + \frac{\alpha}{2\pi} \qquad (2.15)$$

$$g_e \approx 2.001161715 \qquad (2.16)$$

Additionally,

$$(g-2)_{Theoretical} \approx 0.00223183620(86) \qquad (2.17)$$

The (86) is the uncertainty.

$$(g-2)_{Experimental} \approx 0.00223184122(82) \qquad (2.18)$$

$$\text{Difference} = 0.0000000050(12) \tag{2.19}$$

$$\text{A standard deviation of } 4.2\ \sigma \tag{2.20}$$

There is a deviation in the orientation of the magnetic moment of the muon as it spins.

For e^+e^- electron-positron annihilations:

$$(g-2) = 1369080 \times 10^{-11} \tag{2.21}$$

$$(g-2)_{SM} = 14150(110) \times 10^{-11} \tag{2.22}$$

$$SM = \text{Standard Model}$$

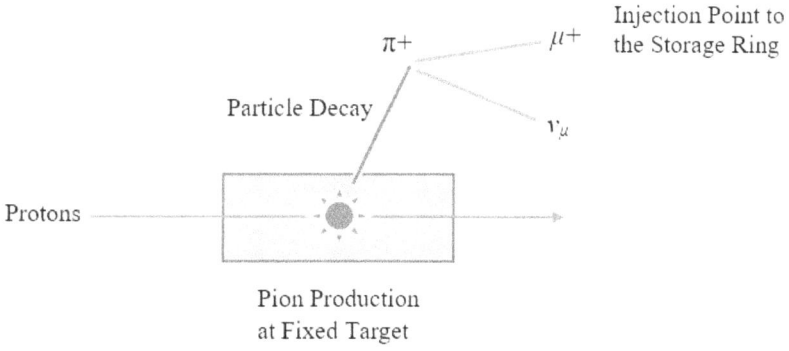

Figure 5. An Illustration of Pion Production from Protons.

Let us now use our current understanding of the g–factor and its related definitions to theorize a potential relativistic equation for the electron's magnetic dipole moment in simpler terms.

Let us illustrate the simple model of an electron moving in a circular orbit of radius "r" with a speed of "v" around the z-axis. The spin magnetic moment "μ_{orb}" will be given by the resulting current times the area of the circle.

$$\vec{\mu}_{orb} = i\vec{A} = \frac{-e}{2\pi r/v}\left(\pi r^2 \vec{a}_A\right) = -\frac{e}{2m_e}\left(\vec{r} \times m_e \vec{v}\right) = -\frac{e}{2m_e}\vec{L}_{orb} \quad (2.23)$$

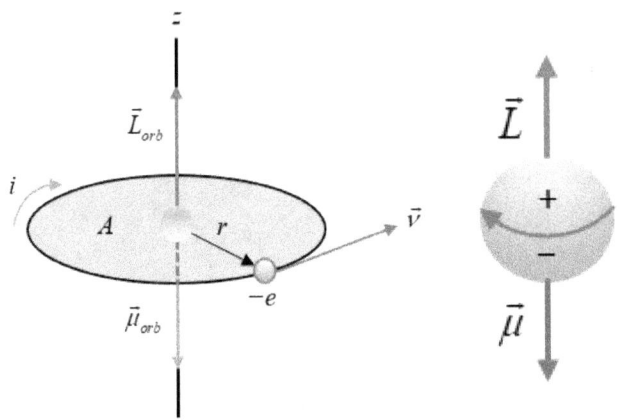

Figure 6. An Atom in a Magnetic Field.

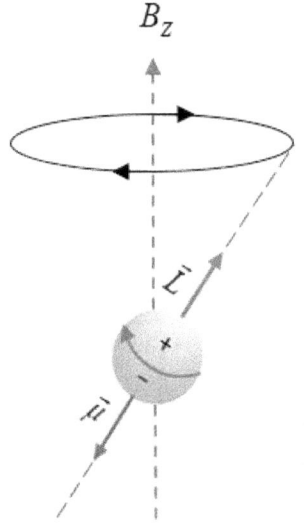

Figure 7. An Electron in a Magnetic Field.

$$\mu_B = \frac{e}{2m_e h} \Rightarrow \text{Bohr Magneton} \quad (2.24)$$

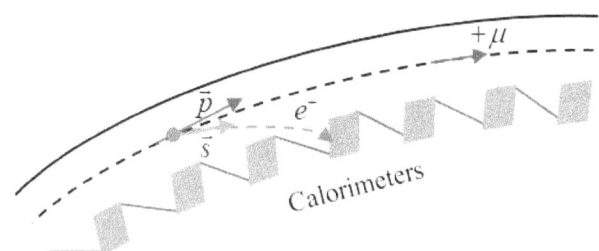

Figure 8. An Illustration of the Storage Ring showing Calorimeters and Particles.

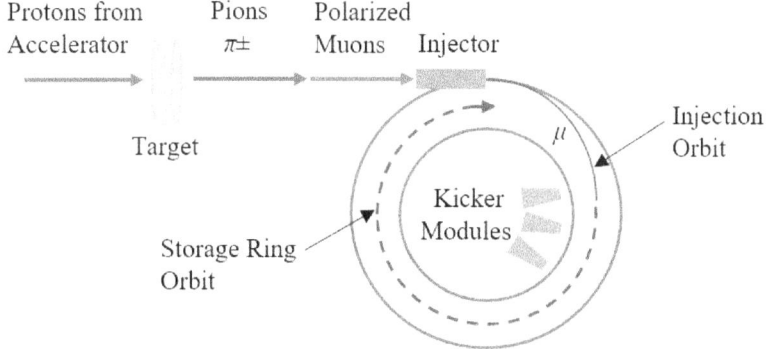

Figure 9. An Illustration of the Storage Ring Devices and Muon Orbits to Measure the Anomalous Magnetic Moment of the Muon.

The electron's non-relativistic magnetic dipole moment, or non-relativistic spin magnetic moment, is given by

$$\mu = g \cdot \frac{-e}{2m_e} L_e \qquad (2.25)$$

$-9.284764620(57) \times 10^{-24} \approx 2.001161715 \cdot (-4.639687313 \times 10^{-24})$

$$L_e \approx \frac{2m_e \mu}{-eg} \approx \frac{2(9.10938356 \times 10^{-31} \, Kg)(-9.284764620 \times 10^{-24} \, J/T)}{(-1.60217662 \times 10^{-19} \, C)(2.001161715)} \qquad (2.26)$$

$$L_e \approx \frac{2m_e\mu}{-eg} \approx 5.275909136 \times 10^{-35} \frac{Kg \cdot m^2}{s} \qquad (2.27)$$

The g–factor value immediately follows from the ratio of a non-relativistic magnetic dipole moment "μ" or "L_e", and a relativistic spin angular momenta "μ'" or "L'_e", which can be both attributed to a spinning electron of known rest mass, and the fine structure constant "α". Aside, for a classical, non-relativistic particle, a typical value of "g" equal to 1, may be used.

$$g \approx \frac{L_e}{L'_e} + \frac{\alpha}{2\pi} \approx \sqrt{1 - \frac{v^2}{c^2}} + \frac{\alpha}{2\pi} \qquad (2.28)$$

The electron's relativistic magnetic dipole moment, μ', also called the relativistic spin magnetic moment, may be denoted as:

$$\mu' \approx \left(\sqrt{1 - \frac{v^2}{c^2}} + \frac{\alpha}{2\pi}\right) \frac{-e}{2m_e} L_e \qquad (2.29)$$

$$\mu' \approx \frac{-e}{2m'_e} L_e + \frac{-e\alpha}{4\pi m_e} L_e \qquad (2.30)$$

The electron's g–factor is about 2, m_e is the rest mass of the electron, L_e is the spin angular momentum ($m_e v r$) of the electron (with magnitude $\hbar/2$ for a Dirac particle as a single particle wave equation), $-e$ is the electron's charge, and "μ" is the electron's magnetic dipole moment or spin magnetic moment. The most accurate value of the electron's non-relativistic magnetic moment "μ" is approximately $-9.284764620(57) \times 10^{-24}$ J/T or A·m² or m³. (NIST, 2021) The electron magnetic moment has been measured to an accuracy of 7.6 parts in 10^{13}. The magnetic moment of an electron is caused by its intrinsic properties of spin and electric charge.

Consequently, according to the Lorentz factor table, $\gamma = 2.000$ exactly, as the velocity approaches the speed of light, if and only if $v/c \approx 0.866025404$, or $\sqrt{3}/2$, and that yields the relativistic equation for the electron's relativistic magnetic dipole moment where "g" is about 2. This effect springs from the fact that "g" is defined as a mixed ratio of a non-relativistic and a relativistic spin angular momenta "L/L'". Aside, $1/\gamma = 0.5$.

Hence, it would be interesting to consider the Lorentz factor using the clock ratio of the frequency "f_{clock}" and the frequency "$f_{calibration}$" for future muon experiments, $\sqrt{1 - f_{clock}^2 / f_{calib.}^2}$.

$$\mu'(f) \approx \left(\sqrt{1 - \frac{f_{clock}^2}{f_{calib.}^2}} + \frac{\alpha}{2\pi} \right) \frac{-e}{2m_\mu} L_\mu \qquad (2.31)$$

The general formula for the magnetic dipole moment "μ" of a classical electron may be expressed as

$$\mu = \int_{V_1}^{V_2} dV = \int_{S_1}^{S_2} I ds = \frac{E}{\rho_\phi} \qquad (2.32)$$

where "V" is a volume that is equivalent to a current "I" times an area, "E" is the magnetic moment energy, and "ρ_ϕ" is the magnetic field flux density, Wb/m^2.

The volume is spatial, and the master clock measurement is temporal, with the master clock unblinding factor, f_{clock}, equal to the relativistic time of the muon as measured in the earth's gravitational field. Therefore, it is possible to suggest that the difference in magnetic dipole moment may be attributed to the relativistic temporal difference between the electron and the muon in six-dimensional space-time. As spatial volume expands and time dilates at a slower rate on the unit of charge due to the larger relativistic mass of the muon per the General Theory of Relativity, the strength of the magnetic dipole field of the muon becomes more concentrated per unit of area of the volume than the electron magnetic dipole field which has the same charge $-e$, less relativistic mass, and a faster rate of spatiotemporal expansion. So, it seems as if the magnetic dipole field of the muon has strengthened in comparison, when in fact is due to the divergence of space-time under the four known forces of nature.

$$\frac{\mu_{muon}}{\mu_e} \approx \frac{V_{muon}}{V_e} \qquad (2.33)$$

The current equations for a muon involve very complex calculations, and some corrections that depend on mass allowances, and bigger contributions for the muon. Hadronic components dominate theoretical uncertainties. If the discrepancy between the theory and the experimental result persists, it can point to new physics. Moreover, the difference "Δa_μ" of the discrepancy constraints very tightly any new physics models, and that has significant implications to interpret any new phenomena.

For the anomalous magnetic moment "$a_\mu(SM)$", we have the following contributions:

$$a_\mu(SM) = a_\mu(QED) + a_\mu(hadronic) + a_\mu(electroweak) \qquad (2.34)$$

$\qquad\qquad\quad(\sim 0.1\%) \quad (\sim 0.00001\%) \quad (\sim 0.0000001\%)$

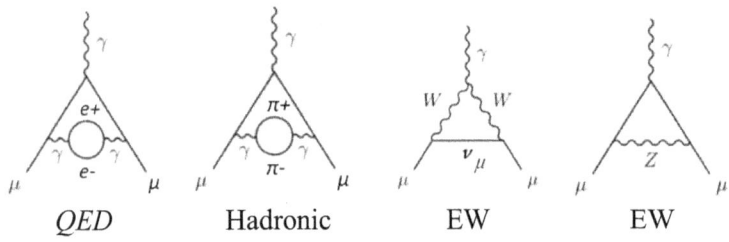

Figure 10. The Contributions to the Anomalous Magnetic Moment.

The measurement of $e^\pm$ and $\mu^\pm$ magnetic dipole moments has been an important benchmark for the development of QED and the standard model of particle physics. Presently, the difference between measured "a_μ" and the prediction of the standard-model is (3.6 σ). Let us propose an approximation for the g–factor that includes the anomalous magnetic moment.

$$g \approx \sqrt{1 - \frac{v^2}{c^2}} + a_\mu(QED) + a_\mu(hadronic) + a_\mu(weak) \qquad (2.35)$$

Task/Result	Anomalous Magnetic Moment $(a_e \equiv g_e / 2 - 1)$
Experiment	$a_{e(EXP)} = 1159652180.73\ (0.28) \times 10^{-12}$
Theory	$a_{e(SM)} = 1159652181.643\ (0.77) \times 10^{-12}$
Difference	$a_{e(EXP)} - a_{e(SM)} = -0.91\ (0.82) \times 10^{-12}$
Deviation	$(1.1\ \sigma)$

Table 1. A Comparison of the Experimental Value and the Theoretical Value of the Anomalous Magnetic Moment of the Electron.

Let us apply the above equations with the following assumptions, $\gamma \approx g - \alpha/2\pi \approx 2.002331846 - 0.001161715 \approx 2.001170131$, as the velocity approaches the speed of light, if and only if $v/c \approx 0.866194133$ for a muon, to obtain the spin angular momentum of a muon using the Fermilab experimental value for the g–factor. (Cohen et Alia, 1987)

Physical Property	Value
Mass (m_μ)	$1.883531627 \times 10^{-28}\ Kg$
Lifetime (τ_μ)	$2.19714(7)\ \mu s$
Charge (q)	$-e$
Intrinsic Spin	$½\ h/2\pi$
Magnetic Moment (μ_μ)	$-4.4904514(15) \times 10^{-26}\ J/T$
Spin g–factor (g_μ)	$2.0023318408(11)$
Gyromagnetic Ratio $(g_\mu \mu_\mu/h)$	$135.69682(5)\ MHz/T$

Table 2. The Physical Properties of Muons.

$$\mu = g \cdot \frac{-e}{2m_\mu} L \qquad (2.36)$$

$$\mu' \approx \left(\sqrt{1 - \frac{v^2}{c^2}} + \frac{\alpha}{2\pi} \right) \frac{-e}{2m_\mu} L \qquad (2.37)$$

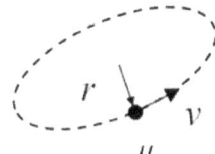

Figure 11. The Orbit of a Muon.

$$-4.4904514(15) \times 10^{-26} J/T \approx 2.0023318408(11) \cdot \frac{-1.602176634 \times 10^{-19} C}{2(1.883531627 \times 10^{-28} Kg)} L_\mu \quad (2.38)$$

$$L_\mu \approx \frac{2m_\mu \mu}{-eg} \approx \frac{2(1.883531627 \times 10^{-28} Kg)(-4.4904514(15) \times 10^{-26} J/T)}{(-1.60217662 \times 10^{-19} C)(2.0023318408(11))} \quad (2.39)$$

$$L_\mu \approx \frac{2m_\mu \mu}{-eg} \approx 5.272862801 \times 10^{-35} \frac{Kg \cdot m^2}{s} \quad (2.40)$$

The g–factor of the muon from the Fermilab experiment in 2021 is 2.0023318408(11) according to NIST CODATA values of the fundamental constants.

The difference between the spin angular momenta of the muon and the electron is 0.003046335 $Kg \cdot m^2/s$, and the difference between the g–factors is 0.001170125, a difference that is near the value that includes the first term "$\alpha/2\pi$" of the a(QED) radiative corrections of perturbation theory.

So, the ratio of the spin angular momenta to g–factors is as follows:

$$\frac{\Delta L}{g - factors} \approx 2.603426984 \quad (2.41)$$

The ratio of the differences of the spin angular momenta and the g–factors is about 260.3% that may be due, but not limited to, the mass difference between electrons and muons, the reciprocal relation between a spin angular momentum and its corresponding g–factor, and the non-linear effect of the relativistic angular motion.

Is the g–factor of an electron = 2.0011614?

g – 2 = the g–factor without taking into account the interactions for a virtual particle?

Is the Standard Model of physics wobbling or is it our misunderstanding of the nature of space-time?

Particle	Symbol	g–factor	Relative Standard Uncertainty
Electron	g_e	−2.00231930436256(35)	1.7×10^{-13}
Muon (Brookhaven Experiment 2006)	g_μ	−2.0023318418(13)	6.3×10^{-10}
Muon (Fermilab Experiment 2021)	g_μ	−2.0023318408(11)	5.4×10^{-10}
Muon (World Experimental Average 2021)	g_μ	−2.00233184121(82)	4.1×10^{-10}
Muon (Theory, June 2020)	g_μ	−2.00233183620(86)	4.3×10^{-10}
Neutron	g_n	−3.82608545(90)	2.4×10^{-7}
Proton	g_p	+5.5856946893(16)	2.9×10^{-10}

Table 3. NIST CODATA Recommended g–factor values.

The electron g–factor is one of the most precisely measured values in physics.

As the Muon spins and becomes magnetic, the magnetic moment of the Muon is about 200 times smaller than the non-relativistic magnetic moment of an electron, and way more sensitive than other particles that may be moving around it in the quantum realm. The Muon acts like a little dipole magnet. If you move the Muon in a circular trajectory its magnetic moment would wobble like a spinning top. Then, you can measure the frequency of the wobble. The measurement would tell you the strength of the magnetic moment of the Muon, the Muon would give off a photon and re-absorb it, which creates a correction to the predicted value for the magnetic moment. That photon can shapeshift into an electron-positron pair, which can annihilate and become another photon which in turn can be re-absorbed. These quantum corrections can affect the value of the magnetic moment of a particle.

The measurement of the g–factor by the Brookhaven National Lab experiment in Long Island, NY, (more than 20 years ago) of the muon deviated from the prediction (slightly above 2) of the current Standard Model and nothing could account for the difference. Researchers found a g–factor, a measurement of the magnetic moment, that was above the predicted value by nearly 3 standard deviations, the g–factor result was the comparison of the measured wobble to the magnetic field with a precision of 0.14 parts per million. A second experiment that was performed by Fermilab in 2017, using a fifty-foot particle storage ring at the Fermi National

Accelerator Laboratory (Fermilab) outside of Batavia, Illinois, confirmed the previous result.

Either the electron or the muon may approach the speed of light, but not achieve light speed due to the fact that they are tardyons, they have rest mass. So, their velocity will typically be less than the speed of light in a particle storage ring like the one at Fermilab.

Speed (units of c), $\beta = v/c$	Lorentz factor, γ	Reciprocal, $1/\gamma$
0.000	1.000	1.000
0.050	1.001	0.999
0.100	1.005	0.995
0.150	1.011	0.989
0.200	1.021	0.980
0.250	1.033	0.968
0.300	1.048	0.954
0.400	1.091	0.917
0.500	1.155	0.866
0.600	1.250	0.800
0.700	1.400	0.714
0.750	1.512	0.661
0.800	1.667	0.600
0.866	2.000	0.500
0.900	2.294	0.436
0.990	7.089	0.141
0.999	22.366	0.045
0.99995	100.00	0.010

Table 4. The Lorentz Factor and its Ratios.

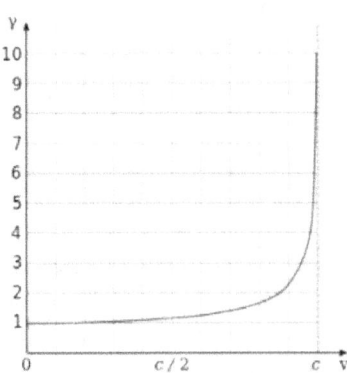

Figure 12. Lorentz Factor "γ" as a Function of Velocity.

In the table above, the left-hand column shows speeds as different fractions of the speed of light (i.e. in units of c). The middle column shows the corresponding Lorentz factor, the right-hand column is the reciprocal. Values in bold are exact.

The Lorentz factor:

$$\gamma = \frac{1}{\sqrt{1-\frac{v^2}{c^2}}} = \frac{1}{\frac{d\tau}{dt}} \qquad (2.42)$$

$$\frac{1}{\gamma} = \sqrt{1-\frac{v^2}{c^2}} = \frac{d\tau}{dt} \qquad (2.43)$$

A blind analysis, as used in a particle physics measurement, is a measurement which is performed without looking at the answer. Blind analyses are the optimal way to reduce or eliminate experimenter's bias, the unintended biasing of a result in a particular direction.

Source of Prediction	Value
Standard Model Prediction	2.00233183<u>19</u>
Initial Brookhaven Result	2.00233184<u>04</u>
International Group Prediction	2.00233183<u>62</u>
Budapest-Marseille-Wuppertal (BMW) Group Calculation (From QCD using a supercomputer)	2.0023318<u>3908</u>

Table 5. The "g – 2" Blinding Numbers.

The effects of Quantum Chromodynamics are present in the existing gluons, and as a result in the existing quarks, from the eighth decimal place of each value onwards. A process where a muon or an electron may be more massive, creating additional particles in the fluctuations of the vacuum, such as a photon and a virtual photon that may decay into a quark-antiquark pair that combines with a photon, and gets absorbed by a muon. Hence, the observation of the magnetic moment differed from the theoretical expectation from the eighth decimal place of each value onwards.

The muon exists for approximately a millionth of a second in the fifteen-meter, or fifty-feet, particle storage ring, going around several hundreds of times before decaying to an electron. Those electrons from the decaying muon parent particles, are detected coming from around the ring. The energy and direction of an electron into a detector around the ring indicate the wobble of the muon parent particle that would specify the muon magnetic moment. The g-value is difficult to calculate. There are virtual processors for hadron computations that are very expeditious to calculate the g-value for quarks that also need to be consistent with the experimental observation. If the computational result is withing a five sigma difference between theoretical and experimental g-value, it would be considered a discovery. So, a computational result of 4.2 sigma, or standard deviations, is very close to the experimental g-value. The contributions of hadrons to the magnetic moment are difficult to compute even for an acceptable value of the standard prediction from a realistic observational result.

A very recent measurement of the anomalous magnetic moment of a positive muon has been 0.46 ppm. The magnetic anomaly of a positive muon can be determined from the precision measurements of two angular frequencies in a muon g – 2 experiment as expressed by $a_\mu \equiv (g_\mu - 2)/2$. The frequency of the difference "ω_a" between the cyclotron frequencies and the precession of spin is incorporated in the intensity of the variation of positrons of high frequency in polarized muons in a magnetic storage ring. The magnetic field of a storage ring may be measured by nuclear magnetic resonance probes. These probes are typically calibrated to the equivalent spin precession frequency of a proton "$\tilde{\omega}'_p$" in a spherical water receptacle at 94.5° F (34.7° C). The ratio of the frequency of the difference and the frequency of the precession of spin of a proton, and other known physical constants, yields the value of a_μ = $\omega_a/\tilde{\omega}'_p(T) = 116592040(54) \times 10^{-11}$ (0.46 ppm). This result would be 3.3 sigma, higher than the prediction of the current standard model, but it agrees with previous measurement. The latest experimental average of $a_\mu = 116592061(41) \times 10^{-11}$ (0.35 ppm) expands the extent between the experimental and theoretical value to 4.2 sigma after unifying previous measurements of both μ^+ and μ^-. The following "g – 2" graph is based on the value of $a_\mu \times 10^9 - 1165900$.

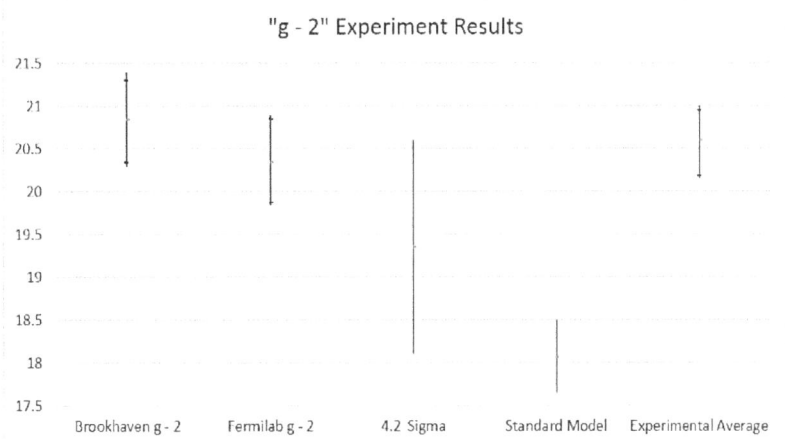

Figure 13. A Graph of the Difference between the theoretical and the most recent measurements of a_μ. The inner tick marks indicate the statistical contribution to the total uncertainties.

The above figure shows the tension between recent current standard model predictions of a_μ together with the measurement performed at Brookhaven from years 1997 to 2001 and the latest measurement at E989 from 2018. The Fermilab result is the most precise measurement of the anomalous magnetic moment of the muon.

The theory has become more rigorous while the experiment has become more effective, but the result of 4.2 sigma still needs to be improved to be grouped to other discoveries in particle physics. This result is attracting greater interest in the scientific community and it has potential for the emergence of new physics. The crucial formula has the key frequency of the unblinding factor of the frequency clock to arrive at an agreeable g-value to previous experiments.

As particles constantly appear and disappear from existence in our physical reality, they may affect more massive particles like muons that are more sensitive to these particles.

Hence, it is probable that the muons are being tilted a little more, causing the muon to have a slightly larger internal magnetic field than normal.

The ratio $R'_\mu(T)$ can be conceptually written in terms of measured quantities and corrections as

$$R'_\mu(T) \approx \frac{f_{clock}\,\omega_a^m\,(1+C_e+C_p+C_{ml}+C_{pa})}{f_{calib}\,\langle\tilde{\omega}'_p(T)(x,y,\phi)\times M(x,y,\phi)\rangle(1+B_k+B_q)} \quad (2.44)$$

The numerator includes the master clock unblinding factor "f_{clock}" or correction for blinding clock offset, the frequency "ω_a^m" of the measured precession of the muon spin relative to the momentum rotation in the magnetic field, and the four beam-dynamics corrections: "C_e" is the "ω_a" for electric field correction, "C_p" is the "ω_a" for pitch correction of vertical beam oscillations, "C_{ml}" is the "ω_a" for the muon loss correction, and "C_{pa}" is the "ω_a" for phase acceptance correction.

The equivalent frequency for the precession of spin for a proton "$\tilde{\omega}'_p(T)$" is anatomized into the calibration procedure for the absolute nuclear magnetic resonance indicated by the magnetic field probes calibration frequency "f_{calib}" and the fields maps, which are weighted by the positrons that are detected, "$\omega'_p(T)(x,y,\phi)$" is the measured shielded proton spin precession frequency map in the storage ring, "$M(x,y,\phi)$" is the muon beam distribution, which is averaged over several timescales given by "$\langle\omega'_p(T)(x,y,\phi)\times M(x,y,\phi)\rangle$". In addition, two fast magnetic transients: "B_k" is the "$\tilde{\omega}'_p(T)$" for the kicker eddy fields correction, "B_q" is the "$\tilde{\omega}'_p(T)$" for the electric quadrupoles transient field correction, that are synchronized to the injection to correct the result. "T" is the water sample temperature at 94.5^0 F (34.7^0 C).

Where have the rest of the Muons gone?

The Large Hadron Collider beauty experiment at CERN investigates the subtle differences between matter and antimatter through the study a particle called the "beauty quark", or "b quark". There are ample quarks created before decay into other particles.

The Large Hadron Collider experiment employs several subdetectors to detect forward particles that are hurled by a one-directional collision. The subdetectors are installed and positioned at a distance from one another along twenty meters from the collision point. The beauty quarks may be tracked and identified by adjustable detectors close to the paths of the circling beam of the Large Hadron Collider.

The Large Hadron Collider beauty experiment at CERN is a very accurate way to test the current standard model. Protons are fired in opposite directions to collide them near the adjustable detectors, to detect the decay of beauty mesons. A meson is a particular combination of a quark and an anti-quark. A beauty meson contains a beauty quark or bottom quark. A beauty quark is associated with the top quark. The most common combination of quarks is the up and down quarks. The beauty quark decays rapidly and is more uncommon. The beauty quarks may decay into strange quarks which may decay further into two electrons, or two leptons. A kaon is a meson having a mass several times that of a pion. A kaon may decay into a muon-antimuon pair, $K^+ \to \mu^-$ and μ^+, or an electron-positron pair, $K^+ \to e^-$ and e^+.

The beauty quark decayed into muons at a slower rate than into electrons, even though the coupling between those two particles is the same, with all other things being equal. Thus, there is a preference toward producing electrons. The decay process produced eighty per cent electrons and twenty per cent muons. The current standard model predicts that the process should produce fifty per cent leptons and fifty percent muons. The speculation about the decay process was that there should likely be another particle that would compensate for this result such as a boson, a very massive Z^0 particle, or a lepton quark.

If time passes at a slower rate on the surface area of the muon, the surface area of its charges may not expand as much as an electron's, which would leave the muon's magnetic field slightly stronger that the electron's emergent magnetic field per unit of square area after expansion. The larger mass of the muon should result in greater relativistic effect around it than around the electron's lower mass. Consequently, there would be lower spatiotemporal expansion around the muon. However, the rate of time is faster around the electron and its decay rate may be higher by a factor of about 4. Hence, some rhetorical questions that arise from these concepts for future research are: What is the effect of space-time on the g-value of the muon? How would the experiment result be if done at very low gravity? What about doing the experiment at sea-level and at a very high altitude to find the relativistic effect difference? How much more is the muon's magnetic field stronger than the electron's magnetic field? What is the effect of doing the ring experiment vertically instead of horizontally? It is hoped that some experimenter, or enquirer, with the proper resources may succeed in conducting the physical experiments that validate the principles and theories of space-time that have been presented.

§ 3. Could all the energies in our universe be added up into a single Lagrangian equation?

The six-dimensional Gluon Standard Model Lagrangian equation of "all there is" may be approximately represented as,

$$\mathcal{L} \equiv \left\{ -\left(\frac{1}{n}\right) trace\, R_{\mu\nu} R^{\mu\nu} + \psi^*(r,t)(iD_e)\gamma^e \psi(r,t) + h.c. \right\} \quad (3.1)$$

$$+ \left\{ \psi_i V^{ij} \psi_j \phi + h.c. \right\} + \left\{ |D_\mu \phi|^2 - V(\phi) \right\} + \left\{ \frac{1}{2} m_P \vec{g} \ell_P - U(\vec{g}) + h.c. \right\}$$

The Lagrangian ≡ { − The Forces of Interaction Particles + The Interaction of Matter Particles with Forces − The Gluon Self-Interactions + (h.c. if needed)}

+ {The Mass for Matter Particles + The Mass for Antimatter Particles + h.c.}

+ {The Mass for the Forces of Interaction Particles − The Higgs Self-Interactions}

+ {The Mass of the Gravitational Forces of Interaction Particles − The Gravitational Self-Interactions + (h.c. if needed)}

The Lagrangian equation represents the approximate sum of all energies in our universe from a relativistic point of view. The gluon standard model is a quantum field theory for six-dimensional space-time.

The first term consists of two matrices. The electromagnetic field strength tensor is represented by "R", but in the context of this equation, the term represents all the ways that all the force carrying particles (bosons) interact with each other. The Higgs is not included in this term. The indices represent a (3 + 3) spatiotemporal formalism. Three spatial dimensions and three conjugate temporal dimensions.

If the term was fully expanded to show the interactions of the individual bosons, it would look like:

The force carrying particles → $L_{gauge} \equiv$ −The field of electromagnetic (bosons) interactions. force matrices −The field of the weak force matrices − The field of the strong force matrices

$$-\frac{1}{n}R_{\mu\nu}R^{\mu\nu} \rightarrow L_{gauge\ bosons} = -\frac{1}{n}F_{\mu\nu}F^{\mu\nu} - \frac{1}{n}W^b_{\mu\nu}W^{b\mu\nu} - \frac{1}{n}S^b_{\mu\nu}S^{b\mu\nu} \quad (3.2)$$

The "b" represents the presence of the three weak force bosons W^+, W^-, and Z^0, and takes into account the nine distinct color gluons which includes the photon in the background of six-dimensional space-time. There is no "b" in the electromagnetic force "F" matrices because photons do not interact with each other unlike $W^\pm$, Z^0, and gluons; even though there is a triadic photonic field. (Nieves, 2021)

The first and second parts gives the Quantum Chromodynamic Lagrangian density for quarks and their gluon field, where the trace represents the six-dimensional matrix $(R_{\mu\nu}R^{\mu\nu})$, "D_e" is the quark covariant derivative, and "γ^ε" are the six-dimensional gamma matrices.

The second term, $\psi^*(r,t)(iD_e)\gamma^e\psi(r,t)$, describes how interaction particles (bosons) interact with matter particles (fermions), or the gauge boson fields and their interactions with the fermionic fields. The fields "ψ" are functions of space and time and also describe antiquarks and antileptons. The asterisk over means that the corresponding vector must be transposed and complex-conjugated; a mathematical device to ensure that the Lagrangian density remains scalar and real.

$$\psi^*(r,t)(iD_e)\gamma^e\psi(r,t) \rightarrow L_{Fermions} = \sum_{Quarks} i\bar{q}\gamma^e D_e q \quad (3.3)$$

$$+ \sum_{Lepton\ "l"} i\bar{\psi}_l \gamma^e D_e \psi_l + \sum_{Lepton\ "r"} i\bar{\psi}_r \gamma^e D_e \psi_r$$

The term, $(iD_e)\gamma^e$, represents how quarks interact with the electromagnetic force, weak force, and the strong force, and the sum over the six distinct quarks of the gluon standard model. The next terms represent the sum over the left-handed and the right-handed leptons, and the term, $(iD_e)\gamma^e$, represents the coupling forces for leptons. The "D_e" is the so-called covariant derivative. It is also used to represent a wave function in classical quantum mechanics. Although this is related to the field representation being used here, the two are not exactly the same.

The "h.c." term represents the Hermitian conjugate of terms 2, 3, and 5. The Hermitian conjugate is necessary if arithmetic operations on matrices produce complex-valued disturbances. By adding h.c. such disturbances cancel each other out; hence, the Lagrangian remains a real-valued function.

The Hermitian conjugate represents the interactions of anti-matter particles with the Higgs field, which is added to the fourth term, or it may represent gravitational interactions, or disturbances, for particles of mass in the fifth term.

The third term, $\psi_i V^{ij} \psi_j \phi + h.c.$, describes how matter or anti-matter particles couple to the Brout–Englert–Higgs field "ϕ", whereby they obtain mass. The entries of the Yukawa matrix "V^{ij}" represent the coupling parameters to the Brout–Englert–Higgs field, which are directly related to the mass of the particle under consideration. These parameters are not theoretically predicted, but have been determined experimentally.

The full Yukawa Lagrangian may be expressed as,

$$L_{Yukawa} = -\bar{q}'_l Y_d \phi d'_r - q'_l Y_u \bar{\phi} u'_r - \bar{L}_l Y_\phi \ell_r \quad (3.4)$$

The first designation are down-type quarks, that is, the down, the strange and the bottom quark, and how they couple to the Higgs field. The second designation are the up-type quarks, the up, the charm, and the top quark, and their coupling to the Higgs field. The third designation are the leptons, "L" and "ℓ" and how they couple to the Higgs field. Quarks and leptons are shown as left-handed "l" or right-handed "r".

The fourth term, $D_\mu \phi^\dagger D^\mu \phi$, or $|D_\mu \phi|^2$, describes how the interaction particles couple to the Brout–Englert–Higgs field. This applies only to the interaction particles of the weak interaction, whereby they obtain their mass. The terms, $D_\mu D^\mu$, are the interaction particles, or force carrying particles, and the term, $\phi^\dagger \phi$, is the Brout–Englert–Higgs field. This has been proven experimentally, because couplings of $W^\pm$ bosons to Higgs bosons have already been verified. Photons do not obtain mass by the Higgs mechanism, whereas gluons are massless because they do not couple to the Brout–Englert–Higgs field. The Lagrangian of the BEH field is given by
$L_{BEH\ Field} = |D_\mu \phi|^2 - V(\phi)$.

The "$-V(\phi)$" term describes the potential of the Brout–Englert–Higgs field. The "V" is the Higgs potential, and "ϕ" is the Higgs field. The "$-V(\phi)$" term also describes how Higgs bosons couple to each other, or how the Brout–Englert–Higgs interacts with itself. Opposite to the other quantum fields, the potential "$-V(\phi)$" has an infinite set of different minima, but not a single minimum at zero. That characteristic makes the Brout–Englert–Higgs field inherently distinct, and leads to spontaneous symmetry breaking, when choosing one of the minima. Hence, matter particles and interaction particles couple differently to this background field, and as a result, they obtain their respective masses.

The fifth term $L_{gravitation} \equiv \left\{ \frac{1}{2} m_P \vec{g} \ell_P - U(\vec{g}) + h.c. \right\}$ describes how particles interact with the gravitational field, and the potential of the gravitational field of all particles of mass at a Planck scale, as explained in the "Quantum Theory of Color Strings." (Nieves, 2022)

The six-dimensional Gluon Standard Model Lagrangian in a highly simplified formulation:

$$\mathcal{L} = L_{gauge\ bosons} + L_{Fermions} + L_{BEH\ Field} + L_{Yukawa} + L_{gravitation} \qquad (3.5)$$

The six-dimensional "$\rho_\mathcal{L}$" stands for the Lagrangian density, which is the density of the Lagrangian function in a differential volume element. In other words, "$\rho_\mathcal{L}$" is defined such that the Lagrangian is the integral over the six-dimensional space-time of the density: $\rho_\mathcal{L} \equiv \left(\int dx^6 \cdot \sqrt{-g} \right) \cdot \mathcal{L}$. Lagrangian mechanics was introduced by the eminent mathematician Giuseppe Luigi Lagrange in 1788 as a reformulation of classical mechanics. Lagrangian mechanics allow the description of the dynamics of a given classical system using only one scalar function, $L = T - V$, where "T" is the kinetic energy and "V" the potential energy of the system. The Lagrangian is used together with the principle of least action to obtain the equations of motion of a system in a very graceful way. The Lagrangian density describes the kinematics and dynamics of a quantum system when handling quantum fields, instead of the discrete particles of classical mechanics.

The Lagrangian density equation is the equation of motion for the Gluon field through six-dimensional space-time, which characterizes the dynamics of the Gluon field strength. The "$\sqrt{-g}$" is the square root of a positive scalar that corresponds to the characteristics of the curvature of the interval, "g" is the determinant of the six-dimensional metric tensor, and "$\vec{g}$" is the gravitational field.

The "n" is the number of spatiotemporal dimensions, for four dimensions, $n = 4$, because the three temporal dimensions are folded, but $n = 6$ for a (3 + 3) formalism with three spatial dimensions and three temporal dimensions.

The six-dimensional Lagrangian density is given by

$$\rho_{Lagrangian} \equiv \left(\int dx^6 \cdot \sqrt{-g} \right) \cdot \mathcal{L} \quad (3.6)$$

$$\rho_\mathcal{L} \equiv \left(\int dx^6 \cdot \sqrt{-g} \right) \left\langle \left\{ -\left(\frac{1}{n}\right) trace\ R_{\varepsilon\beta} R^{\varepsilon\beta} \right\} \right. \quad (3.7)$$

$$+ \left\{ \psi^*(r,t)(iD_e)\gamma^e \psi(r,t) + h.c. \right\}$$

$$+ \left\{ \psi_i V^{ij} \psi_j \phi + h.c. \right\} + \left\{ |D_\mu \phi|^2 - V(\phi) \right\} + \left. \left\{ \frac{1}{2} m_P \vec{g} \ell_P - U(\vec{g}) + h.c. \right\} \right\rangle$$

The Lagrangian Energy Density ≡ The integral over space-time of the density × The Lagrangian

The Specific Lagrangian Gravitational Pressure ≡ The Specific Lagrangian Energy Density

$$\left(\int dx^6 \cdot \sqrt{-g} \right) \left\{ \frac{1}{2} m_P \vec{g} \ell_P - U(\vec{g}) + h.c. \right\} \equiv \rho_{S\mathcal{L}} \quad (3.8)$$

Let us consider a novel mathematical framework of quantum field theory which is a combination of classical field theory, special relativity, and quantum mechanics. The letter "Z" designates the six-dimensional Path Integral Formulation of Quantum Mechanics to sum up the quantum and classical fields over a spatiotemporal path with length "r" that may be nonlinear and multidimensional. The term "$e^{i\mathcal{L}}$" is Euler's equation that endows each field of the path integral with its wave property. The following path integral formulation evolved and drew its inspiration from the original four-dimensional path integral that was furthered by the eminent physicist Richard Feynman. (Feynman, 1948)

$$k_{field} = \frac{2\pi}{\lambda_{field}} \quad (3.9)$$

$$\frac{2\pi}{\lambda_{field}} < \lambda_{UV\ Cutoff} \quad (3.10)$$

$$Z \equiv \int_{\frac{2\pi}{\lambda_{field}} < \lambda_{UV\ Cutoff}} \left| \vec{\mathfrak{R}} \cdot \left(\vec{\psi} \cdot \left(-\vec{F} \cdot -\vec{W} \cdot -\vec{S} \right) \cdot \vec{\phi} \cdot \vec{g} \right) \right| \cdot e^{i\mathcal{L}} dr \quad (3.11)$$

$$Z \equiv \int_{k_{field} < \lambda_{UV\ Cutoff}} \left| \vec{\mathfrak{R}} \cdot \left(i^2 \cdot \vec{\psi} \cdot \vec{F} \cdot \vec{W} \cdot \vec{S} \cdot \vec{\phi} \cdot \vec{g} \right) \right| \cdot e^{i\mathcal{L}} dr \quad (3.12)$$

$$Z \equiv \int_{k_{field} < \lambda_{UV\ Cutoff}} \left| \prod_{n=1}^{6} \left(\vec{\mathfrak{R}} \cdot \vec{\mathbb{F}}_n \right) \right| \cdot e^{i\mathcal{L}} dr \quad (3.13)$$

The path integral formulation indicates the quantum amplitude through a transition from one specific field configuration to another, formulated as a summation over all the possible paths that could connect them. A configuration is a distinct value for every field, at every spatiotemporal point. In other words, a path integral adds up all of the infinitesimal field contributions of all the ways the fields could evolve from start to finish.

Where "k_{field}" is the mode number of a distinct mode in a field, $k_{field} = 2\pi/\lambda_{field}$. The *UV cutoff* is the ultraviolet cutoff, that represents the wavelength "$\lambda_{UV\ Cutoff}$" at which a solvent absorbance in a 1 *cm* path length cell is equal to 1 AU (absorbance unit) using water in the reference cell, which is an arbitrary minimum cutoff of energy, momentum, and length. Each field may be a combination of modes, each mode constitutes an oscillation with a distinct wavelength. In this context, the above notation limits the field configurations in the path integral to those that do not oscillate too vigorously, or limits the field configurations to weak field or low-energy situations.

The term "$\vec{\mathfrak{R}}$" is the six-dimensional Robertonian operator to represent the infinitesimal field contributions that are going to be added up in our path integral. (Nieves, 2020) The term "D_e" represents the covariant derivative. The square root "$\sqrt{-g}$" is the square root of a positive scalar that corresponds to the characteristics of the curvature of the interval, that is, a scalar related to gravitation. The letters "$\vec{F}, \vec{W}, \vec{S}$" represent all the other bosonic force fields such as photons, gluons, $W^{\pm}$ and Z^0 bosons and how they interact with each other, the Greek letter "Ψ" represents fermions such as leptons and quarks, and the Greek letter "ϕ" represents the Brout–Englert–Higgs field that gives mass through spontaneous symmetry breaking.

Furthermore, the path integral equation is the amplitude, angular frequency, and phase at a particular pressure point and configuration of the path integral for undergoing a transition to another pressure point and configuration. If an action is complex, the number of the action may be expressed as a magnitude and a phase angle. The magnitude may be expressed as the peak value of the wave, or a root mean square value, which is a real number, of a periodic wave for comparative purposes. A field is composed of harmonics of its fundamental frequency. In our context of quantum particles, the high-energy harmonics have small wavelengths and high frequencies.

All the world lines of a particle in the bulk pass through the unobserved quantum particle. Once the particular particle is observed from one or more universe(s), the wave function collapses to that particular universe or those particular universe(s), sharing the particle in question under the energy conservation law of the multiverse. The observed particle pinches the medium of the remaining world lines of its collapsed wave function.

Hence, the reversibility of a path integral, or of the Schrodinger's equation, is not violated unless the wave function collapses. Consequently, if the wave function collapses, the particle-wave(s) follow their world line(s) under the applicable laws of thermodynamics, other natural laws, and the emergent direction of space and time. Therefore, a particle may co-exist at different spatial locations or time periods in distinct but related world lines of sibling universes. A particle may be a part of distinct particle systems in different universes, or may exist independently, during the same time period.

The gravitational strength, "g", is Newtonian for a weak field, but Einsteinian, where $g = c^2/r$, for a strong field, like inside a black hole or near a Big Bang singularity. The gravitational charge of any particle is the spatiotemporal pressure around it and throughout its inner space-time or plenum, which may originate from external gravitational fields, gravitons, and/or gravitational waves. All particles are theorized to have a gluonic substructure even fermions, or particles that do not interact with some other particles such as photons, neutrinos, or leptons, according to the Gluon Standard Model. The universal gravitational charge represents energy density, and spatiotemporal energy density is equivalent to spatiotemporal pressure.

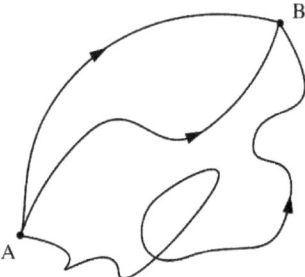

Figure 14. An Illustration of Path Integrals Between Two Spatiotemporal Points.

In conclusion, the action is the seemingly random wiggle or motion of a quantum field or wave along the path of least action in six-dimensional space-time. The product of the quantum field contributions represents the amplitude of the waveform. The action is the exponent of the kernel of spatiotemporal growth as the medium is modulated by the forces of matter, energy, and the emergence of space-time. The exponential function of the product of the imaginary number "i" and the Lagrangian is the resultant path integral of the quantum mechanical system. The Lagrangian density equation represents the six-dimensional path integral of the spatiotemporal pressure points along the trajectory of the action of the quantum mechanical system from start to finish.

Chapter 7

The Accretion or Dissolution of Mass

§ 1. What is the source and the sink of all color strings?

The colorness of the color string can be denoted as

$$\sqrt{k} \equiv \mp i \left(\sqrt{\frac{m'_s v_s^2}{m'_s c^2}} \right) e^{(\ell_s + i\omega_s T_s)\frac{1}{2}} \tag{1.1}$$

$$k \equiv \pm \left(\frac{v_s^2}{c^2} \right) e^{\ell_s + i\omega_s T_s} \tag{1.2}$$

$$\pm k = \frac{v_s^2}{c^2} e^{\ell_s} \left(\cos \omega_s T_s + i \sin \omega_s T_s \right) \tag{1.3}$$

$$\omega_s = 2\pi f_s \tag{1.4}$$

where $\pm k$ is the colorness energy of the string, ω_s is the associated angular frequency of the color string, f_s is the linear frequency, ℓ_s is the length of the color string, T_s is the period of the color string waveform, and m'_s is the relativistic mass of the color string.

$$m_s \equiv \frac{m_0}{\sqrt{1 - \left(\frac{\ell_s \omega_s}{\ell_s \omega_c} \right)^2}} \equiv \frac{m_0}{\sqrt{1 - \left(\frac{\omega_s}{\omega_c} \right)^2}} \tag{1.5}$$

$$m'_s \equiv \frac{m_s}{\sqrt{1 - \left(\frac{v_s}{c} \right)^2}} \tag{1.6}$$

where m_0 is the rest mass of the color string, ω_c is the angular frequency of light, v_s is the translational velocity of the color string

179

through space, the term $1/\sqrt{1-(\omega_s/\omega_c)^2}$ is the relativistic angular frequency factor of the color string mass, and $\vec{g}$ is the gluonic field between colors and anti-colors in *Newtons / ±k*.

The energy of a color string is given by

$$Energy = \pm k\vec{g}\ell_s \qquad (1.7)$$

The manifestation of mass and energy comes from the spatiotemporal medium and its quintessential substances of space and time. It is theorized that as the spatiotemporal wavelets interfere between points of source, there may be shear and torsion between the spatiotemporal wavelets as they expand or contract at any arbitrary spatiotemporal point, causing the production of color strings that may expand, contract, twist, or be static, depending on their angular frequency. The spatiotemporal pressure of an element of color string of length "ℓ_s" would be equivalent to its energy density which would manifest the property of relativistic mass "m'_s" for a color string if the temporal angular frequency is slightly less than the angular frequency of light. This relativistic mass "m'_s" may also be referred to as the rest mass "m_0" if the oscillating color string is not translating through space.

§ 2. How does a color string sustain its energy density?

It is theorized that a color string, or a color graviton, can sustain its relativistic mass if its angular frequency remains higher than the angular frequency of the dissolution threshold of its medium. If its angular frequency decreases below the angular frequency of the dissolution threshold of its medium, the color string relativistic mass, or the color graviton relativistic mass, would no longer be sustained and would begin to dissolve back to its medium.

As the angular frequency of the color string stays higher than the threshold of dissolution, the color string, or the color graviton, sustains its relativistic mass through the kinetic angular momentum process impacted by the expanding wavelets around the color string or color graviton. The temporal wavelets are the prime movers of the color strings or color gravitons as they expand around the energy density volume to transfer angular momentum to the relativistic masses of the color strings or color gravitons.

Let us consider a moving point *P(x, y, 0)* on the surface of a string as it moves along a circle of radius "*r*" on the *x-y* plane with the positive *z*-axis out of the page. The circle is centered at the origin "*O*" of the *x-y* plane. The object travels with angular frequency "ω", meaning the angle *θ(t)* formed by the arc of its path with the positive *x*-axis at time "*t*" can be written as

$$\theta(t) = \theta_0 + \omega t \tag{2.1}$$

where $\theta_0 = 0$ at $t = 0$. At *θ(t = 0) = 0*, the point *P(x, y, 0)* is at the position *P(x₀, y₀, z₀) = (r, 0, 0)*, or initially on the positive *x*-axis at a distance "*r*" from the origin "*O*".

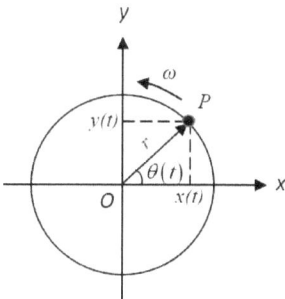

Figure 1. A point *P(x, y, 0)* traveling along a circle of radius "*r*" with angular frequency "ω".

The equations of motion for the point "*P*" are given by

$$x(t) = r \cos\theta(t) = r \cos(\omega t) \tag{2.2}$$

$$y(t) = r \sin\theta(t) = r \sin(\omega t) \tag{2.3}$$

Differentiating the motion equations to obtain the velocity and acceleration equations, we have

$$\frac{dx(t)}{dt} = v_x(t) = -\omega r \sin(\omega t) \qquad (2.4)$$

$$\frac{dv_x(t)}{dt} = a_x(t) = -\omega^2 r \cos(\omega t) \qquad (2.5)$$

$$\frac{dy(t)}{dt} = v_y(t) = \omega r \cos(\omega t) \qquad (2.6)$$

$$\frac{dv_y(t)}{dt} = a_y(t) = -\omega^2 r \sin(\omega t) \qquad (2.7)$$

The vector $\vec{r}(t)$ is the vector connecting the origin "O" to the traveling point "P" at any given instant of time. The velocity vectors $\vec{v}_x(t)$ and $\vec{v}_y(t)$ are always perpendicular to the vector $\vec{r}(t)$. The accelerations $\vec{a}_x(t)$ and $\vec{a}_y(t)$ are called centripetal accelerations, always pointing toward center of the circular path. The velocity $\vec{v}_x(t)$ or $\vec{v}_y(t)$, and the acceleration $\vec{a}_x(t)$ or $\vec{a}_y(t)$, are also always perpendicular to each other. Hence, the velocity of the point "P" is always tangential to the circle, even if the angular frequency "ω" is constant over time.

Substituting for $\sin^2(\omega t) + \cos^2(\omega t) = 1$, for any real value "$\omega t$", we have

$$v(t) = \sqrt{v_x^2(t) + v_y^2(t)} = \sqrt{\omega^2 r^2 \sin^2(\omega t) + \omega^2 r^2 \cos^2(\omega t)} = \omega r \qquad (2.8)$$

The value of $v(t)$ may vary over time since both "ω" and "r" are variables, so the direction of $\vec{v}(t)$ and its magnitude may change with the geometry of the color string relativistic mass.

The effect of the acceleration $\vec{a}(t)$ is then to change the direction of the velocity $\vec{v}(t)$ and its magnitude. The magnitude of a velocity

$\vec{v}(t)$, for circular motion or any other motion, may change over time.

$$\frac{d|\vec{v}(t)|}{dt} = \frac{d}{dt}\sqrt{v_x^2(t)+v_y^2(t)} = \frac{2v_x(t)a_x(t)+2v_y(t)a_y(t)}{2\sqrt{v_x^2(t)+v_y^2(t)}} = \frac{\vec{v}(t)\cdot\vec{a}(t)}{|\vec{v}(t)|} \qquad (2.9)$$

Thus, $\vec{v}(t)$ is perpendicular to $\vec{a}(t)$, the magnitude of $\vec{v}(t)$ may vary and its direction may vary for a nonzero acceleration.

If the point "P" is not always traveling in a circle, the radius "r" may vary over time as r(t) for any value of "t". Hence, the value of the acceleration may also vary as given by

$$a(t) = \omega^2 r(t) = \frac{v^2(t)}{r(t)} \qquad (2.10)$$

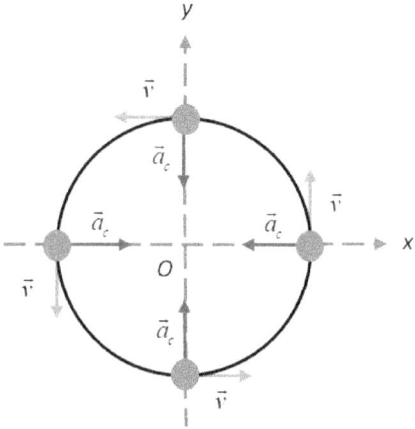

Figure 2. The Velocities and Accelerations around the Circular Path when $\theta = 0^0, 90^0, 180^0,$ and 270^0, using the value of $\theta = \theta(t) = \omega t$.

The centripetal acceleration always points to the center of the circular path while the velocity is always tangential to the circular. The definition of centripetal acceleration is $a_c = v^2/R = \Sigma F_c/m$, where "m" is the moving mass.

In summary, as a point "P" travels along a circular path with angular velocity $\vec{v}(t)$, its acceleration points to the center of the circle while its velocity is tangential to the circle. The motion of the point "P" on the surface of the color string is theorized to be caused by the expanding spatiotemporal wavelets around the color string relativistic mass.

§ 3. What is the quantum charge of a color string?

The equation for the quantum charge of a color string may be written as

$$q_s \equiv 2\pi(\ell_s \cdot t_s) \tag{3.1}$$

$$q_s = \sum_{p=1}^{n} a_p (l_p \cdot t_p) \tag{3.2}$$

where "q_s" is the color string quantum charge in Coulombs, the Planck quantum charge is equal to the product of the Planck length "l_p" and the Planck time "t_p", "a_p" is the dimensional coefficient of the geometry at or around a point, and "n" is the number of quantum charges on a color string.

During previous research, it was explicated that a Planck quantum charge is equal to $8.713036182 \times 10^{-79}$ $m \cdot s$ or Coulomb, a quantum of charge that is infinitesimal with respect to the estimated magnitude of the length of a color string, $\sim 10^{-35}$ m. If an atom could be magnified to the size of our solar system, then a color string could be magnified to the size of a golden cane palm tree. (Nieves, 2020)

§ 4. What makes the gluonic field stronger as color string charges separate? How do color string charges separate or spring back?

As a quantum force field tries to separate color string charges, it is theorized that the spatiotemporal medium may be stretched as the frequencies of color string masses increase, raising the strong nuclear force of the inherent stress and torsional energy of each color string which makes it harder to pull the medium apart. The color strings would drift toward a lower spatiotemporal pressure, concentrating their high energy of colorness in the low pressure regions.

The spatiotemporal medium tends to contract to counterbalance its varied colorness density and strain, to spring back to its previous uniform spatiotemporal state. It is possible to suggest that the medium in this process may also exhibit spatiotemporal elasticity. As the spatiotemporal medium contracts, there are spatiotemporal regions of greater pressure from the wavelet interference which tend to spring back to smooth out the volume of the spatial deformation. The strong nuclear force is attractive as the medium is stretched, but it becomes repulsive at a distance of less than approximately 0.7 x 10^{-15} meters, as the medium springs back into a more uniform spatiotemporal state at its core. The range of a strong nuclear force is very short and it is assumed identical for all color strings. It is assumed that one possible way to separate these color strings may be to create a state of matter known as color string-gluon-quark plasma.

The equation for the strong nuclear force for a color string may be written as

$$F_{SN} \equiv -m'_s \ddot{a}_s e^{\frac{-1}{d(d-2r)}} \qquad (4.1)$$

or in terms of the reduced Planck constant, we have

$$F_{SN} \equiv -\hbar c e^{\frac{-1}{d(d-2r)}} \qquad (4.2)$$

where "d" is the distance between the fundamental surfaces of two color strings, "r" is the average radius of two color strings, "$-\ddot{a}_s$" is the deceleration of the volume of a color string, "$\hbar$" is the reduced Planck constant, "c" is the speed of light, and "$\hbar c$" is a physical constant with dimensions of $(Kg \cdot m^3)/s^2$, or a volumetric force.

The negative sign in the exponential gives the interaction a finite effective range, and the strong nuclear force strengthens as the distance "d" increases, where $d > 2r$. It is interesting to consider the geometrical representation of the exponent as the curvature of the spatiotemporal medium between two color strings, $-1/m^2$, under the same conditions of $d > 2r$. Hence, that would make the strong nuclear force equal to $-\hbar c e^{-R}$, where "R" is the trace of the Ricci curvature tensor, or Ricci scalar.

Thus, it is relevant to note that color strings at great distances will hardly interact any longer, as interaction forces fall off exponentially with increasing the distance "d" or separation. Consequently, the strong nuclear force would counterbalance the decrease of interaction to spring back the color strings to a distance of greater interaction.

Hence, the Coulomb potential for the strong nuclear force field between two color strings with identical charge, that decreases more rapidly with distance, may be written as

$$V_s \equiv -\frac{q_s^2}{4\pi\varepsilon_r d}e^{-d(d-2r)} \equiv -\frac{\ell_s^2 \cdot t_s^2}{4\pi\varepsilon_r d}e^{-d(d-2r)} \qquad (4.3)$$

where "q_s" is the charge of a color string in Coulombs, "t_s" is a temporal period, and "ε_r" is the relative permittivity of the spatiotemporal medium. (Yukawa, 1935)

Hypothetically, the exponent for the Coulomb potential may be written in terms of the reciprocal of the curvature of the spatiotemporal medium between two color strings, under the previous condition, as follows:

$$V_s \equiv -\frac{q_s^2}{4\pi\varepsilon_r d}e^{-\frac{1}{R}} \equiv -\frac{\ell_s^2 \cdot t_s^2}{4\pi\varepsilon_r d}e^{-\frac{1}{R}} \qquad (4.4)$$

or in terms of the reduced Planck constant, we have

$$V_s \equiv -\hbar c e^{-\frac{1}{R}} \qquad (4.5)$$

The strong nuclear force is approximately 137 times more powerful than the electromagnetic force. An estimate was obtained for the value of $m_s' \ddot{a}_s$ based upon the energy required to dissociate two color strings.

$$F_{snf} = \frac{F_{emf}}{\varepsilon_r} = 137 \cdot F_{emf} \qquad (4.6)$$

where $\varepsilon_r = \varepsilon_{actual}/\varepsilon_0$ is the relative permittivity of the electron with respect to the Planck charge "q_p", approximately 1/137, in free space-time. (Nieves, 2020)

§ 5. What is the angular momentum of a color string?

Let us consider a cross-section of the color string with a system of two identical rotating point particles, each particle has half the relativistic mass $m'_s/2$ of the color string, moving in a circle of radius "R", $180°$ out of phase $(\vec{r}_1 = -\vec{r}_2)$, at an angular velocity $\vec{\omega} = \omega_s \hat{z}$ in a plane parallel to, but at a spatial distance "h", above the x-y plane of a Cartesian coordinate system.

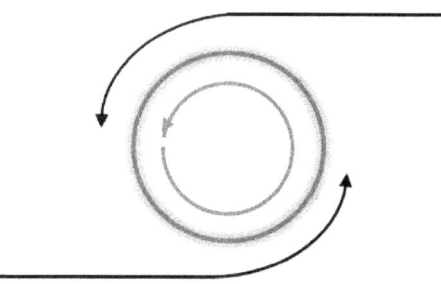

Figure 3. An Illustration of the Angular Momentum of Wavelets on a Color String Cross-Section.

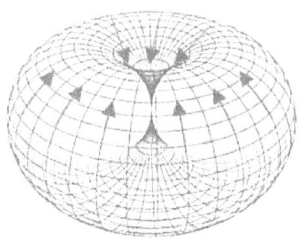

Figure 4. An Illustration of the Angular Momentum of a Color Graviton.

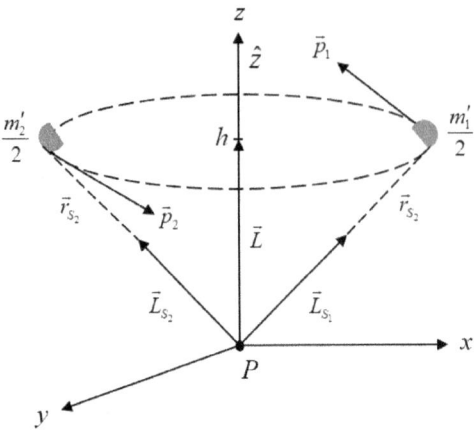

Figure 5. An Illustration of Angular Momentum for A Color String.

The angular momentum of the color string system is $\vec{L}$ for each relativistic mass about a spatiotemporal point "$P(x, y, 0)$". Let us denote the kinematic equations of the system.

$$\vec{L} = \vec{L}_{S_1} + \vec{L}_{S_2} = (\vec{r}_{S_1} \times \vec{p}_1) + (\vec{r}_{S_2} \times \vec{p}_2) \qquad (5.1)$$

$$\vec{L} = \frac{m'_s}{2} R^2 \omega_z \hat{z} - h \frac{m'_s}{2} R\omega_z \hat{r}_1 + \frac{m'_s}{2} R^2 \omega_z \hat{z} - h \frac{m'_s}{2} R\omega_z \hat{r}_2 \qquad (5.2)$$

$$\vec{L} = \frac{m'_s}{2} R^2 \omega_z \hat{z} - h \frac{m'_s}{2} R\omega_z \hat{r}_1 + \frac{m'_s}{2} R^2 \omega_z \hat{z} + h \frac{m'_s}{2} R\omega_z \hat{r}_1 \qquad (5.3)$$

$$\vec{L} = m'_s R^2 \omega_z \hat{z} \qquad (5.4)$$

Therefore, the angular momentum of a color string is a function of its relativistic mass, the radius of its cross-section, and its angular frequency. At the Planck scale, it could be denoted as $\vec{L} = \pm \hbar \hat{z}$.

§ 6. The Interference and the Geometry of a Cluster of Wavelets about a Point.

In the early 17th century, the eminent Mathematician and Astronomer Johannes Kepler stated that the most compacted packing possible of identical spherical objects was the arrangement one observed in a grocer's pyramid of oranges. (Kepler, 2010) Any spherical object in the interior of Kepler's three-dimensional arrangement touches twelve other spherical objects, and the spatial volume filled by the spherical objects equals $\pi/\sqrt{18}$, or approximately 37/50.

The proof of Kepler's claim came four hundred years later. Hence, no other compacted three-dimensional packing of identical spherical objects can be more compacted. (Hales, 2005)

Later, in the 1690s, the eminent physicist Sir Isaac Newton declared that a central spherical object could touch no more than a dozen surrounding identical spherical objects with a constant diameter. Newton's assertion is known as the kissing-number problem, proved to be right in 1953. (Schütte et Alia, 1953)

Therefore, Kepler's pyramid of oranges or Newton's kissing-number problem suggests a similar spatiotemporal wavelet model of identical spherical wavelet packing at the instant of interference, under stress and torsion, with a dozen wavelets converging simultaneously toward a central arbitrary point.

Let us now consider a cluster of spatiotemporal wavelets at the initial point of interference as the wavelets expand or contract simultaneously and interfere with spatiotemporal shear stress and torsion that may produce color strings in the process.

Given a cluster of spatiotemporal wavelets, let us construct the corresponding graph or adjacency matrix, $\left[L_{i,j} \right]$. The adjacency matrix indicates which wavelets are interfering that may produce color strings through spatiotemporal shear stress and torsion.

There are six adjacent wavelets converging at an arbitrary spatiotemporal point, one from each direction of each spatiotemporal dimension in six-dimensional space-time. The center point of the central spatiotemporal wavelet may be an arbitrary point. In our example, there are fifteen potential initial points of interference for the six adjacent wavelets converging into an arbitrary spatiotemporal point, $P(x, y, z)$. (Arkus et Alia, 2009 and 2011)

$$[L_{i \cdot j}] = \begin{vmatrix} a_{11} & a_{12} & a_{13} & a_{14} & a_{15} & a_{16} \\ a_{21} & a_{22} & a_{23} & a_{24} & a_{25} & a_{26} \\ a_{31} & a_{32} & a_{33} & a_{34} & a_{35} & a_{36} \\ a_{41} & a_{42} & a_{43} & a_{44} & a_{45} & a_{46} \\ a_{51} & a_{52} & a_{53} & a_{54} & a_{55} & a_{56} \\ a_{61} & a_{62} & a_{63} & a_{64} & a_{65} & a_{66} \end{vmatrix} = \begin{vmatrix} 0 & 1 & 1 & 1 & 1 & 1 \\ 0 & 0 & 1 & 1 & 1 & 1 \\ 0 & 0 & 0 & 1 & 1 & 1 \\ 0 & 0 & 0 & 0 & 1 & 1 \\ 0 & 0 & 0 & 0 & 0 & 1 \\ 0 & 0 & 0 & 0 & 0 & 0 \end{vmatrix} \quad (6.1)$$

Whereas each element a_{ij} of the adjacency matrix is a binary value, answering the yes-or-no question "Do wavelets "i" and "j" interfere?

All the binary data of the adjacent matrices is contained in the upper triangle of the matrix, which has $n(n-1)/2$ elements. Each element has two possible values, and so the total number of adjacency matrices is $2^{n(n-1)/2}$.

Any valid packing of wavelets has a continuous, unbranched path that threads from one wavelet to the next throughout the overall structure, like a long polymer chain. Could this elemental process be similar to the process of how color strings may get attached into color string polymers during wavelet interference? (Biedl et Alia, 2001)

A value of one in the adjacency matrix designates a pair of unit wavelets whose distance between their centers is exactly unity. Hence, for any $a_{ij} = 1$ the distance is unity, $d_{ij} = 1$. A cluster of wavelets would be feasible only if every distance element satisfies the constraint $d_{ij} \geq 1$. Any distance smaller than one would mean that two wavelets were occupying the same volume before interference.

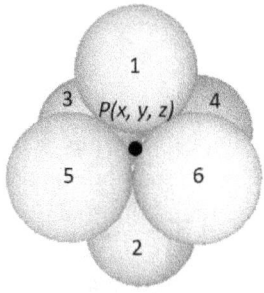

Figure 6. A Cluster of Six Spatiotemporal Waves.

To determine the geometry of a cluster of wavelets, it is necessary to determine the *x, y,* and *z* coordinates of all "*n*" wavelets. There is an Algebra rule that states that "*3n*" equations are needed to determine "*3n*" unknown variables; nonetheless, only *(3n–6)* equations are necessary.

The energy of the cluster of wavelets depends only on the relative positions of the "n" wavelets, not on the orientation or absolute position of the entire cluster of wavelets.

It is possible to arbitrarily assume that one wavelet is at the origin of the Cartesian coordinate system and another is exactly one unit of spatial distance away along the positive *x*-axis. In this manner, the six coordinates become fixed.

Consequently, the *(3n–6)* equations supplied by the values of one in the adjacency matrix are exactly the number needed to locate the rest of the wavelets at the instant of interference.

If there were precisely *(3n–6)* wavelet interferences, and at least three interferences per wavelet, a cluster of wavelets would have a property called minimal wavelet rigidity.

If any wavelet had only one or two interferences, it could waggle or sway freely. Such a cluster of wavelets could not be a *max(C_n)* configuration because the unrestricted wavelet could turn to interfere with at least one more wavelet, thereby increasing C_n.

Each of the *(3n–6)* wavelet equations with a value of $r_n = 1$, may be written as,

$$r_n = \sqrt{(x_i - x_j)^2 + (y_i - y_j)^2 + (z_i - z_j)^2} \qquad (6.2)$$

to denote the spatial distance between the center points of the wavelets "*i*" and "*j*", this family of equations, or these simultaneous equations, need to be solved in order to regain the spatial Cartesian coordinates of all the wavelets.

n	3n – 6	n(n – 1)/2	C_n	Multiplicity
1	–	0	0	1
2	–	1	1	1
3	3	3	3	1
4	6	6	6	1
5	9	10	9	1
6	12	15	12	2
7	15	21	15	5
8	18	28	18	13
9	21	36	21	52
10	24	45	24, 25	259, 3
11	27	55	27, 28, 29	1620, 20, 1

Table 1. A Summary of the Cluster Varieties of Wavelets.

where *"n"* is the number of wavelets, multiplicity is the number of distinct ways, or varieties, to attain the bound cluster of the wavelets, *"C_n"* is the total number of wavelet interference points, the maximum wavelet interference number is *(3n–6)*, and *"n(n–1)/2"* is the number of wavelet interferences to form a cluster. (Hoy et Alia, 2010 and 2012)

The cluster of nine wavelets has a property not observed in any other *max(Cn)* cluster configuration, the highest value of *"C_n"*, up to this number *"n"* of wavelets: it is the property of flexibility. The overall structure can be twisted around one axis by torsion, without breaking any interference bonds between the wavelets. In the flexible *(n = 9)* cluster, two wavelet manifolds joined by a border of interference can stress and twist slightly such as it would be expected for spatiotemporal shear stress and torsion. It is possible to theorize that in the *(n = 9)* cluster of wavelets, there is probability that each point of interference may produce distinct composite color strings or polymers that may generate all possible combinations of gluons in the Gluon Standard Model.

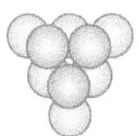

Figure 7. A Cluster of Nine Spatiotemporal Waves.

Therefore, it is possible to theorize that spatiotemporal shear stress "τ_s" and torsion may generate numerous color strings through spatiotemporal wave interference in a variety of physical processes. This phenomena may explain the appearance of virtual particles in free space through the process of chromosynthesis of color strings that may produce color gravitons, gluons, bosons, quarks, hadrons and elementary particles. It is hypothesized that this process could be bidirectional, so that as particles decay, and eventually disintegrate back to its color string elements, these elements of energy may revert to the quintessential substances of space and time as their angular frequency decreases to the surrounding spatiotemporal frequency of their medium.

Spatiotemporal torsion from the interference of spatiotemporal wavelets, like a linear force, will produce both stress and strain. However, unlike linear stress and strain, spatiotemporal torsion may cause a twisting stress, called spatiotemporal shear stress "τ_s", and a rotation, called spatiotemporal shear strain "γ_s".

Spatiotemporal Shear Stress "τ_s" $\equiv$ *Pressure* $\equiv$ *Energy Density*

The torsional shear stress "τ_s" is given by

$$\tau_s = \frac{Tr}{J} \tag{6.3}$$

where "T" is the applied torque, $E/\omega t$, "J" is the polar second moment of inertia, $\pi \rho^4 / 2$, that depends only on the spatiotemporal geometry, and "r" is the radius.

The twist angle "ϕ" starts at zero and increases linearly as a function of "ℓ_s". On the other hand, the change of angle "γ_s" is constant along the length of the color string.

The spatiotemporal shear strain can be written as

$$\gamma_s = \frac{\rho \phi}{\ell_s} \tag{6.4}$$

Hence, the length of a color string may be denoted in terms of shear strain and torsion,

$$\ell_s = \frac{\rho \phi}{\gamma_s} \tag{6.5}$$

where "γ_s" is spatiotemporal shear strain or the angle of rotation, "ρ" is the radius of the cross section of the color string, "ϕ" is the angle of torsion, and "ℓ_s" is the length of a color string.

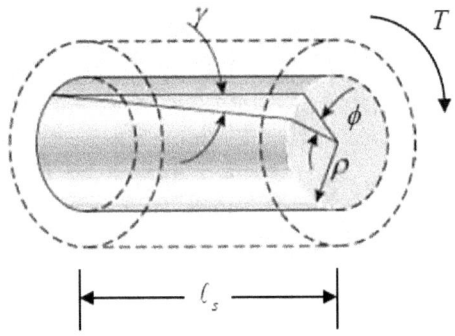

Figure 8. An Illustration of the Shear Strain of a Color String.

Let us represent the spatiotemporal shear stress equation in matrix form for a system of six spatiotemporal wavelets.

$$\{\tau_{i \cdot j}\} = \{P_{i \cdot j}\}[L_{i \cdot j}] \tag{6.6}$$

$$\{P_{i \cdot j}\} = \begin{vmatrix} P_{t_x t_x} & P_{t_x t_y} & P_{t_x t_z} & P_{t_x x} & P_{t_x y} & P_{t_x z} \\ P_{t_y t_x} & P_{t_y t_y} & P_{t_y t_z} & P_{t_y x} & P_{t_y y} & P_{t_y z} \\ P_{t_z t_x} & P_{t_z t_y} & P_{t_z t_z} & P_{t_z x} & P_{t_z y} & P_{t_z z} \\ P_{x t_x} & P_{x t_y} & P_{x t_z} & P_{xx} & P_{xy} & P_{xz} \\ P_{y t_x} & P_{y t_y} & P_{y t_z} & P_{yx} & P_{yy} & P_{yz} \\ P_{z t_x} & P_{z t_y} & P_{z t_z} & P_{zx} & P_{zy} & P_{zz} \end{vmatrix} \tag{6.7}$$

$$[L_{i \cdot j}] = \begin{vmatrix} 0 & 1 & 1 & 1 & 1 & 1 \\ 0 & 0 & 1 & 1 & 1 & 1 \\ 0 & 0 & 0 & 1 & 1 & 1 \\ 0 & 0 & 0 & 0 & 1 & 1 \\ 0 & 0 & 0 & 0 & 0 & 1 \\ 0 & 0 & 0 & 0 & 0 & 0 \end{vmatrix} \qquad (6.8)$$

$$\{\tau_{i \cdot j}\} = \begin{vmatrix} \tau_{t_x t_x} & \tau_{t_x t_y} & \tau_{t_x t_z} & \tau_{t_x x} & \tau_{t_x y} & \tau_{t_x z} \\ \tau_{t_y t_x} & \tau_{t_y t_y} & \tau_{t_y t_z} & \tau_{t_y x} & \tau_{t_y y} & \tau_{t_y z} \\ \tau_{t_z t_x} & \tau_{t_z t_y} & \tau_{t_z t_z} & \tau_{t_z x} & \tau_{t_z y} & \tau_{t_z z} \\ \tau_{x t_x} & \tau_{x t_y} & \tau_{x t_z} & \tau_{xx} & \tau_{xy} & \tau_{xz} \\ \tau_{y t_x} & \tau_{y t_y} & \tau_{y t_z} & \tau_{yx} & \tau_{yy} & \tau_{yz} \\ \tau_{z t_x} & \tau_{z t_y} & \tau_{z t_z} & \tau_{zx} & \tau_{zy} & \tau_{zz} \end{vmatrix} \qquad (6.9)$$

where $\{\tau_{i \cdot j}\}$ is a spatiotemporal shear stress tensor, $\{P_{i \cdot j}\}$ is a pressure tensor, or energy density tensor, in *Newtons/m²*, at every point of interference for all wavelets, and $[L_{i \cdot j}]$ is a constant adjacency matrix for ($n = 6$) number of wavelets expanding through space.

In terms of spatiotemporal curvature, the spatiotemporal shear stress equation may be denoted as,

$$\{\tau_{i \cdot j}\}[L_{i \cdot j}] = \frac{c^4}{8\pi G}\{R_{i \cdot j}\}[L_{i \cdot j}] \qquad (6.10)$$

Lowering indices with g^{ij} and dividing by "L" we have,

$$\tau = \frac{c^4}{8\pi G}R = \frac{R}{\kappa} \qquad (6.11)$$

where $\{R_{i \cdot j}\}$ is the Ricci curvature tensor of the six-dimensional Einstein Field Equations and "R" is its trace or the scalar curvature determined by the intrinsic geometry of the manifold near a given point, $\{\tau_{i\cdot j}\}$ is the six-dimensional spatiotemporal shear stress tensor and "τ" is its trace, and "κ" is the Einstein gravitational constant, 2.077 x 10^{-43} N^{-1}.

$$R_{i\cdot j} = \begin{vmatrix} R_{t_x t_x} & R_{t_x t_y} & R_{t_x t_z} & R_{t_x x} & R_{t_x y} & R_{t_x z} \\ R_{t_y t_x} & R_{t_y t_y} & R_{t_y t_z} & R_{t_y x} & R_{t_y y} & R_{t_y z} \\ R_{t_z t_x} & R_{t_z t_y} & R_{t_z t_z} & R_{t_z x} & R_{t_z y} & R_{t_z z} \\ R_{xt_x} & R_{xt_y} & R_{xt_z} & R_{xx} & R_{xy} & R_{xz} \\ R_{yt_x} & R_{yt_y} & R_{yt_z} & R_{yx} & R_{yy} & R_{yz} \\ R_{zt_x} & R_{zt_y} & R_{zt_z} & R_{zx} & R_{zy} & R_{zz} \end{vmatrix} \quad (6.12)$$

The Ricci curvature tensor can be characterized by measurement of how a shape like a sphere is deformed as the sphere moves along geodesics in the space. The Ricci curvature is the geometrical object that controls the growth rate of the volume of metric sphere in a manifold.

In the case of the trace-free Ricci tensor, given by $R_{ij} = (2\Lambda/n - 2)g_{ij}$, where "$n$" is the number of dimensions, "g_{ij}" is the metric tensor, and "Λ" is the cosmological constant, the trace-free Ricci tensor is proportional to the metric, because this condition is equivalent to saying that the metric is a solution of the vacuum Einstein Field Equations with a cosmological constant.

§ 7. What is the color charge of a color string?

The color charge of a color string may be visualized as the set of inherent properties such as angular frequency, laterality, and quantum spin of an element of color energy and its associated relativistic mass. In color string theory, spin is understood by the rotation of the string around its axis.

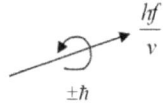

Figure 9. An Illustration of a Color String with Left Laterality.

where is "$\pm hf/v$" the conserved linear momentum, "$\pm\hbar$" is the angular momentum with left or right laterality. Discrete quanta may be stacked to form longer strings with their momentum vectors aligned.

A color string may consist of a discrete quantum, or many discrete quanta of a shared frequency that are stacked with aligned momentum vectors, and are phase entangled with each other. A color string has linear momentum that is represented by a vector, angular momentum represented by its axial rotation about its linear momentum vector, and laterality or direction of rotation. The motion of the string depends on both the linear momentum and the direction of laterality, because of how a color string may emit or absorb a quantum or quanta of colorness. The energy of the quantum consists of, but it is not limited to, the axial angular momentum, the linear momentum, and its colorness, the higher these frequencies are, the greater the energy of the quantum. The laterality of the color string is the angular polarity while its linear direction is its linear polarity. The overall polarity may be defined using Fleming's right or left hand rules.

Furthermore, a geometric feature of the color string is its infinitesimal spatiotemporal volume. The more energy a color string has, or the higher its angular frequency, the smaller its cross-section becomes, and the less energy the color string has, the larger its cross section. If several color strings are stacked with aligned momentum vectors into a composite color string, then as the angular frequency of the composite color string increases, its volume decreases, and consequently, the spatiotemporal volume of the composite color string is less. Hence, the spatiotemporal volume is a function of angular frequency. Similar to an accomplished figure skater who goes into a spin, as he pulls his arms in, he spins faster, and as he extends his arms, he spins slower. Consequently, the above concepts beg the rhetorical question, at low energies, $E \ll 1/\pm k\vec{g}\ell_s$, would the open color string behave like a spin-1 gauge boson with relativistic mass on the world volume of a color D-brane? Concludingly, there is a lacuna in our knowledge of new physics that follow from our findings and conclusions, that would benefit from further research, including experimental evaluation to extend and further test the concepts and quantum theory of color strings that have been developed and presented.

References

Arkani-Hamed, Nima and Dimopoulos, Savas. (2004) *Supersymmetric Unification Without Low Energy Supersymmetry And Signatures for Fine-Tuning at the LHC.* Retrieved from arXiv:hep-th/0405159v2 on 5-10-2021.

Arkus, N., V. N. Manoharan and M. P. Brenner. (2009). *Minimal energy clusters of hard spheres with short range attractions.* Physical Review Letters 103:118303.

Arkus, N., V. N. Manoharan and M. P. Brenner. (2011). *Deriving finite sphere packings.* SIAM Journal on Discrete Mathematics 25(4):1860–1901.

Ashtekar, Abhay. (1986) *New variables for classical and quantum gravity.* Physical Review Letters. 57 (18): 2244–2247.

Biedl, T. E., et Alia. (2001). *Locked and unlocked polygonal chains in three dimensions.* Discrete and Computational Geometry 26:269–281.

Bohm, David. (1952). *A Suggested Interpretation of the Quantum Theory in Terms of 'Hidden Variables' I.* Physical Review. 85 (2): 166–179.

Born, M and Wolf, E. (1999). *Principles of Optics: Electromagnetic Theory of Propagation, Interference and Diffraction of Light* (7th edition). Cambridge University Press.

Cartan, Élie. (1981) [1938], *The theory of spinors*, New York: Dover Publications, ISBN 978-0-486-64070-9.

CERN. (1999) *Particle Physics Education CD-ROM.*

C.N. Yang and T.D. Lee. (1952) *Statistical Theory of Equations of State and Phase Transitions. I. Theory of Condensation.* Phys. Rev. 87: 404-409.

C.N. Yang and T.D. Lee. (1952) *Statistical Theory of Equations of State and Phase Transitions. II. Lattice Gas and Ising Model.* Phys. Rev. 87: 410-419.

Cohen, E. Richard, Taylor eds, Barry N., Res, J. (1987) *The 1986 CODATA Recommended Values of Fundamental Physical Constants*, National Bureau of Standards, 92(2), 1.

de Boer, W. (1994) *Grand Unified Theories and Supersymmetry in Particle Physics and Cosmology.* Institut fur Experimentelle Kernphysik Universitat, Karlsruhe, Germany.

de Broglie, L. (1927). *La mécanique ondulatoire et la structure atomique de la matière et du rayonnement.* Journal de Physique et le Radium. 8 (5): 225–241.

de Broglie, L. (1967). *Le Mouvement Brownien d'une Particule Dans Son Onde.* C. R. Acad. Sci. B264: 1041.

De Carvalho, Vanuildo S. and Freire, Hermann. (2013) *Breakdown of Fermi liquid behavior near the hot spots in a two-dimensional model: A two-loop renormalization group analysis.* Nuclear Physics. B (Print), v. 875, p. 738-756.

Feynman, R. P. (1948). *Space-Time Approach to Non-Relativistic Quantum Mechanics.* Reviews of Modern Physics. 20 (2): 367–387.

Hales, T. C. (2005). *A proof of the Kepler conjecture.* Annals of Mathematics 162:1065–1185.

Hoy, R. S., J. Harwayne-Gidansky and C. S. O'Hern. (2012). *Structure of finite sphere packings via exact enumeration: Implications for colloidal crystal nucleation.* Physical Review E 85:051403.

Hoy, R. S., and C. S. O'Hern. (2010). *Minimal energy packings and collapse of sticky tangent hard-sphere polymers.* Physical Review Letters 105:068001.

Kepler, J. (2010). *The Six-Cornered Snowflake: A New Year's Gift.* Philadelphia: Paul Dry Books.

Lawrence Livermore National Laboratory. (1995) *Science & Technology Review September 1995.* Figure 1, Page 26.

National Institute of Standards and Technology (NIST). (2021) *The most accurate value of "μ_{em}", and CODATA values of the fundamental constants.* Physics. U.S. Department of Commerce.

Nieves, Robert. (2020) *A Dynamic Theory of Space-Time: A Matter of Waves.* Published by Kindle Direct Publishing, Amazon.com, Inc. ISBN 9798667276289.

Nieves, Robert. (2021) *A Synthesis of Quantum Gravity.* Published by Kindle Direct Publishing, Amazon.com, Inc. ISBN 9798715826565.

Peskin, Michael; Schroeder, Daniel (1995). *An introduction to quantum field theory* (Reprint ed.). Westview Press. ISBN 978-0201503975.

Raychaudhuri, A. K. (1955). *Relativistic cosmology I.* Phys. Rev. 98 (4): 1123–1126.

Selleri, F. and Van der Merwe, A. (1990). *Quantum paradoxes and physical reality.* Kluwer Academic Publishers. pp. 85–86.

Schütte, K., and B. L. van der Waerden. (1953). *Das Problem der dreizehn Kugeln.* Mathematische Annalen 125:325–334.

Wigner, E. P. (1931). *Gruppentheorie und ihre Anwendung auf die Quanten mechanik der Atomspektren.* Braunschweig, Germany: Friedrich Vieweg und Sohn. pp. 251–254.

Yukawa, H. (1935). *On the Interaction of Elementary Particles.* Proc. Phys.-Math. Soc. Jpn. 17 (48).

Zwiebach, Barton. (2009) *A First Course in String Theory.* Second Edition. Cambridge University Press.

www.ingramcontent.com/pod-product-compliance
Lightning Source LLC
Chambersburg PA
CBHW071356210526
45465CB00001B/123